REMINISCENCES OF LOS ALAMOS, 1943–1945

STUDIES IN THE HISTORY
OF MODERN SCIENCE

Editors:

ROBERT S. COHEN, *Boston University*

ERWIN N. HIEBERT, *Harvard University*

EVERETT I. MENDELSOHN, *Harvard University*

VOLUME 5

REMINISCENCES OF LOS ALAMOS
1943—1945

Edited by

LAWRENCE BADASH

Dept. of History, University of California, Santa Barbara, U.S.A.

JOSEPH O. HIRSCHFELDER

Dept. of Chemistry, University of Wisconsin, U.S.A.

and

HERBERT P. BROIDA

Dept. of Physics, University of California, Santa Barbara, U.S.A.

D. REIDEL PUBLISHING COMPANY

DORDRECHT : HOLLAND / BOSTON : U.S.A.

LONDON : ENGLAND

Library of Congress Cataloging in Publication Data
Main entry under title:

Reminiscences of Los Alamos, 1943–1945.

 (Studies in the history of modern science ; v. 5)

  1. Atomic bomb–History–Addresses, essays, lectures.
2. Physicists–United States- Biography–Addresses, essays, lectures.
3. Scientists–United States- Biography--Addresses, essays, lectures.
4. Los Alamos, N. M.–Description–Addresses, essays, lectures.
I. Badash, Lawrence. II. Hirschfelder, Joseph Oakland, 1911–
III. Broida, Herbert P. IV. Series.
QC791.96.R44   623.4'5119   80-36731
ISBN 90-277-1097-X
ISBN 90-277-1098-8 (pbk.)

3-0885-500ts

Published by D. Reidel Publishing Company,
P.O. Box 17, 3300 AA Dordrecht, Holland.

Sold and distributed in the U.S.A. and Canada
by Kluwer Boston Inc., Lincoln Building,
160 Old Derby Street, Hingham, MA 02043, U.S.A.

In all other countries, sold and distributed
by Kluwer Academic Publishers Group,
P.O. Box 322, 3300 AH Dordrecht, Holland.

D. Reidel Publishing Company is a member of the Kluwer Group.

First published in 1980
Reprinted in 1985
Reprinted in 1988

Printed in The Netherlands

In Memory of Herbert P. Broida

Royalties from the sale of this book will be donated to the American Physical Society Fund for the Herbert P. Broida Prize.

TABLE OF CONTENTS

PREFACE

Although the World War II efforts to develop nuclear weapons have inspired a very large literature, it struck us as noteworthy that virtually nothing existed in the form of firsthand accounts. *Now It Can Be Told*, by General Leslie Groves, the Manhattan Project's military commander, is probably the most prominent exception, but the scientists themselves seem to have shown little interest in publishing their reminiscences.

Believing that it would be not only worthwhile for posterity, but extremely interesting for the present generation to hear about the aspirations, fears, and activities of those who participated in this watershed of science and government collaboration, we arranged the public lecture series represented by this book.[1] We chose to focus upon Los Alamos since the project's efforts culminated there. The isolated laboratory in New Mexico was created to design and construct the first atomic bombs. More scientific brainpower was accumulated there than at any time since Isaac Newton dined alone, and the interactions with this community are of sociological interest, as the results of their work are of political import.

The lectures were held weekly during the first months of 1975, on the Santa Barbara campus of the University of California. For their hard work, generous support, and warm encouragement, we wish to thank Merrie Walker, Louise Cannell, Sue Bell, Charles Stidd, Deans Bruce Rickborn and Henry Offen, EG&G, Inc., General Electric TEMPO Center for Advanced Studies, and the UCSB College of Creative Studies, Quantum Institute, and Department of History. Richard Feynman's Santa Barbara talk has appeared already in the California Institute of Technology magazine *Engineering and Science* **39** (Jan.–Feb. 1976), 11–30. We are grateful to Edward Hutchings, Caltech Director of Publications, for permission to use his edited version of this lecture. Portions of Bernice Brode's talk appeared as 'Tales of Los Alamos,' *Los Alamos Scientific Laboratory Community News*, (June–Sept. 1960).

L.B.
J.O.H.
H.P.B.

ix

NOTES

[1] During the planning phase for the lecture series the *Bulletin of the Atomic Scientists* began publishing an interesting series of articles by a number of scientists who worked at the several Manhattan Project locations. These have been edited by Jane Wilson into a book entitled *All in Our Time. The Reminiscences of Twelve Nuclear Pioneers* (Chicago: Educational Foundation for Nuclear Sciences, 1975).

INTRODUCTION

During the 1930s, there was so much talk in the popular press about harnessing the energy of the atom that Ernest Rutherford, the leading nuclear physicist of his day, felt obliged to defuse the conjectures by calling them "moonshine."[1] He was joined in this effort by other scientific luminaries such as Robert A. Millikan[2] and Albert Einstein[3]. After all, they argued, though enormous amounts of energy were released in certain nuclear reactions, far more energy had to be fed into the cyclotrons, Cockcroft–Walton accelerators, and Van de Graaff machines of the day than could be extracted from them. Based upon realistic scientific expectations of beam intensities and efficiencies of reactions, these men were correct in regarding nuclear transformations as highly valuable research phenomena for a better understanding of nature, but not as prototypes of usable energy cornucopias.

That they were wrong is not to their discredit. Scientists may be taken to task when their predictions, based on reasonably complete knowledge, are found to be in error. But they cannot fairly be criticized for failing to predict new phenomena of nature. And it was the discovery of just such a phenomenon that provided the key to utilizing the energy stored in atomic nuclei.

Towards the end of 1938 the great German radiochemist Otto Hahn and his colleague Fritz Strassmann convinced themselves that they had detected barium, an element from the middle of the periodic table, upon bombarding the heaviest element, uranium, with neutrons. Certain beta radioactivities in the material suggested the production of man-made elements heavier than uranium, but Hahn had been at his trade nearly thirty-five years and knew barium when he saw it. This puzzle was interpreted by Hahn's long-time physicist colleague Lise Meitner and her nephew Otto Frisch as the splitting of uranium nuclei into roughly equal pieces upon neutron bombardment, and they gave the name 'fission' to the new process. It was the fission fragments that had yielded the confusing beta decays. Most significant was recognition that the mass of the fission fragments was less than that of the initial ingredients in the reaction, and the energy conversion of this mass difference (by Einstein's $E = mc^2$ equation) was soon confirmed. Each fission event released considerably more energy than the nuclear

xi

L. Badash, J. O. Hirschfelder and H. P. Broida (eds.), Reminiscences of Los Alamos 1943–1945, xi–xxi.

reactions heretofore studied (wherein the energy was largely that of the motion of particles), and millions of times more than any chemical reaction, atom for atom.[4]

Niels Bohr communicated news of Frisch's verification experiments, which were carried out in the former's Copenhagen laboratory, to a large meeting of physicists in Washington, D.C., early in 1939. So electrifying was the announcement that even before Bohr had finished speaking, physicists rushed to telephones to instruct colleagues in their home institutions to repeat the experiments. Within days nuclear fission was widely accepted as a real phenomenon.

There were many questions that quickly came to mind. Was it the abundant, heavy isotope of uranium, U-238, which fissioned, or the lighter isotope, U-235, which was found in uranium minerals to the extent of only 0.7 per cent? Were any neutrons emitted in the fission reaction? If neutrons were emitted, how many in each event, and could they be made to strike other fissionable nuclei, causing a repetition of the sequence — a chain reaction? With what geometry of components, with what purity of materials, with what velocity of neutrons could the reaction likelihood be maximized? And if a chain reaction of fissioning nuclei could be achieved, could it be controlled to serve as an energy generator (i.e., a nuclear reactor), or would the reaction speed uncontrollably to conclusion, releasing its energy almost instantaneously, as in a bomb?

These and other technical matters were widely discussed and investigated during 1939. The subject was not kept secret; quite the opposite, in fact, for about one hundred papers on fission were published in the scientific literature that year, and further items appeared in newspapers and magazines.[5] A measure of fission's notoriety is the fact that at this time scientists in America, England, France, Germany, the Soviet Union, and Japan began work that led to the exploration of military applications of nuclear energy during World War II. But this, of course, suggests the reason why fission was so interesting. Not only was it a new phenomenon, fascinating for its intrinsic scientific significance; it also bore the potential for remarkable applications in a world darkening under the gathering clouds of war.

Despite such famous examples in the history of science as Archimedes' defense of Syracuse, Galileo's occasional contributions to military engineering, and the more organized efforts on both sides to produce poison gases during World War I, scientists had relatively little interaction with their governments. Hitler's dreams of world conquest, however, had heightened their political interests in general, and his racial policies, which resulted in the expulsion of numerous Jewish scholars from Germany, had shattered the isolation of their scientific lives.[6]

While many emigrées settled in other European countries, most came to the United States, where a large population and a more developed concept of public higher education offered more job opportunities. The brilliance contained in this intellectual migration, coupled with a solid, diversified, and recently mature American enterprise, decisively shifted the global 'scientific center of gravity' from Europe to the new world. Hitler's prejudices had not only deprived himself of talent of the highest quality, but had thrust it upon his future enemies.[7]

When Germany's ally, Italy, showed signs of adopting similar racial laws, her most famous physicist, Enrico Fermi, took his Jewish wife and family to America.[8] During the 1930s, Fermi's laboratory in Rome had been a center for the study of neutron-induced reactions. When working with uranium he had gathered the same sort of puzzling evidence that Hahn and his colleagues ultimately had resolved. Now at Columbia University in New York, Fermi studied nuclear fission with the goal of achieving a chain reaction. Another emigrée at Columbia with a similar purpose was Leo Szilard, a Hungarian who overflowed with ideas and was at his best in encouraging others to action.[9]

Although research progressed in many laboratories in the first half of 1939, its scope and intensity were unsatisfactory to Szilard. He felt that government interest and funding were vital. Motivating him was the recognition that many excellent scientists remained in Germany,[10] and that if they provided Hitler with an explosive fission device he most certainly would use it.[11] Since an atomic bomb had the potential to release vastly greater amounts of energy than any chemical explosive, its employment conceivably could alter the course of the war for which Hitler clearly was preparing. Most European universities were state institutions, and their scientists, to a greater degree than their American contemporaries, were civil servants, familiar in dealing with government bureaucracies. Thus it was that Szilard conceived the idea of approaching President Roosevelt directly. With Hungarian colleagues Eugene Wigner and Edward Teller, he got Albert Einstein to write a letter to the president urging support of uranium research. Einstein's fame assured that the president would take the matter seriously and, indeed, modest funding and a governmental connection were initiated towards the end of 1939.[12]

During the next two years there was significant progress — for example, U-235 was shown to be the isotope that fissioned, and sufficient neutrons were seen to be emitted in each fracture of a nucleus to maintain a chain reaction, provided the neutrons were not lost through the surface of the system or captured by impurities — but hopes of constructing either a reactor or a bomb within the anticipated duration of the war fluctuated. More than once it appeared that nuclear efforts should be curtailed and the scientists' talents employed on less 'science fiction' type war projects. On one such occasion in mid-1941, American efforts were rescued by an enthusiastic British report which concluded that sufficient U-235 could be separated to construct a weapon whose critical mass would be smaller than heretofore believed necessary, and which therefore would be of a size that an airplane could deliver.[13] Also during this period more energetic and effective governmental organizations were established to manage science for the war effort, specifically the National Defense Research Committee (NDRC) and then the Office of Scientific Research and Development (OSRD), headed by Vannevar Bush and his deputy, James B. Conant.

With America's entry into the war at the end of 1941, nuclear R&D projects received much more financial support. The goals were seen; the means to these ends were less certain. Clearly, U-235 had to be separated from the 140 times more abundant U-238, with which it was found. But since both are uranium with identical chemistry, only physical techniques, based upon their small mass difference of three units, would suffice. Gaseous diffusion, centrifuge, liquid thermal diffusion, electromagnetic separation, and other methods were pursued, each of them having severe, perhaps fatal, difficulties. Another, even more sensational, approach to a weapon was to build it of an element not found in nature. It was found that U-238 could absorb neutrons to become U-239. This isotope would emit beta particles to become the first man-made element, later called neptunium-239. In turn, neptunium also exhibited beta decay and was transformed into plutonium-239. Like U-235, Pu-239 would fission under neutron bombardment, releasing enormous amounts of energy. Yet the manufacture, atom by atom, of an entirely new element was mind-boggling.

For greater efficiency and protection from possible coastal attacks, most of the research was centralized at the University of Chicago, under the leadership of Arthur H. Compton.[14] It was here, under the stands of the football field, that Fermi built his structures of bricks of graphite (to slow the neutrons to their most efficient velocity), in which pieces of uranium metal and uranium oxide were arranged. As industry provided him with materials of greater and greater purity — far exceeding any standards it had been called upon for before — the closer he came to achieving a self-sustaining chain reaction.

Finally, on 2 December 1942, Fermi removed the control rods of cadmium,

which 'soaked up' neutrons, watched in quiet satisfaction as his instrumentation indicated the generation of an expanding (or exponential) chain reaction,
and then shut down the reaction by reinserting the control rods. Nuclear
energy had been produced and controlled. The event had a double significance. Proof that a chain reaction could be initiated meant that a bomb was
a real possibility. Many scientists, uncomfortable in their role as weapons
designers, had hoped that some physical law would prevent the realization
of a chain reaction. Knowing, however, that it could be achieved, meant that
Hitler's scientists would learn the same thing, and it was clear who must
manufacture the weapon first. The other consequence of Fermi's 'Chicago
Pile-1' was that plutonium was a viable alternative should U-235 encounter
insuperable production obstacles. Reactors have high densities of neutrons
and offer the logical place to bring neutrons in contact with U-238.

Several months earlier the Army Corps of Engineers had been brought
into these fission activities. Brigadier General Leslie Groves was given responsibility for translating the scientists' laboratory experiments and plans into
production plants for uranium-235 and plutonium. An efficient administrator,
whose personality commanded respect though not affection, Groves was
willing to make decisions. He quickly acquired large sites in Tennessee and
Washington, and before long thousands of workmen, using trainloads of
materials, were erecting huge factories and small cities to house the construction people and then those who would run the plants.

At Oak Ridge, Tennessee, the chief structures were the gaseous diffusion
plant, the electromagnetic separation plants, and the largest coal-fired power
plant ever built at one time, which furnished prodigious amounts of electricity. The gaseous diffusion plant consisted of an immense series of leak-tight
containers, each vessel bisected by a porous barrier. Molecules containing
U-235 were three mass units lighter than molecules incorporating U-238, and
were more likely to pass through the holes of the barrier. The gas molecules'
normal, random motion, therefore, created a separation of these isotopes, and
the fraction richer in U-235 was pumped to the next higher stage while the
depleted fraction went to a lower one. This process had to be repeated many
times through the numerous stages before a bomb-grade product, enriched to
over ninety percent U-235, could be extracted from the top of the chain.

As if this wonder of chemical engineering was not impressive enough at
first glance, there were additional problems to solve. The only gaseous compound of uranium is uranium hexafluoride, which is poisonous and extremely
corrosive. Moreover, if allowed to come in contact with water vapor in the
air it would clog the holes in the barrier. So the entire system of containers,

pipes, pumps, etc., had to be constructed of materials that would permit no corrosion, and be built to vacuum-tight standards that industry had never been called upon before to achieve. The barriers had to be thin enough to act as porous membranes, yet rigid enough to sustain construction handling and operating pressure differentials. They had to contain billions of microscopic holes of a uniform size. Production of this component was so difficult that the rest of the system was designed and well under construction before the successful manufacture of this item was achieved.

This illustrates a key feature of the Manhattan Project — so named because much of the early work was done in institutions around New York City, where it was coordinated by the Army's Manhattan Engineer District. Because it was feared that Germany would succeed in building an atomic bomb, and because it was believed that the only way to counter this eventuality was for the Allies to construct one first, no path to this end could be dismissed unless it had been *proven* to be a bad one. This explains what otherwise appears as duplicated effort. The electromagnetic plants at Oak Ridge fall into this category. They consisted of numerous vacuum chambers across which a powerful magnetic field was thrown. Since charged particles are bent into curved paths by such fields, and the radii of curvature depend upon the particles' masses, the two isotopes of uranium should have been separated into different collectors in a single stage, unlike the statistical gaseous diffusion process. In practice, the beams of particles created a space charge which destroyed the neat separation, and the need to open the chambers to remove the collectors meant that much time was lost re-evacuating them. The electromagnetic process did furnish some enriched U-235 at a critical time, but the gaseous diffusion technique was capable of greater production. An interesting sidelight is the circumstance that the magnet coils were wound with Treasury Department silver, since the necessary quantity of copper was unavailable due to other wartime needs. After the war the precious metal was returned to Fort Knox.

At Hanford, Washington, on the banks of the Columbia River, five production reactors were erected along with the necessary chemical separation plants. This too was a vast construction project, and it too was not always smooth sailing. One difficulty was the so-called 'slug crisis.' The reactors were designed with numerous horizontal channels to hold the short cylinders of uranium. After a period of time in the reactor, the cylinders would be extracted from the far side of the structure and eventually taken to the chemical plant, where they would be dissolved in acid and a plutonium compound separated from the other constituents. Unlike Fermi's demonstration pile, these reactors required cooling by the river water, and to prevent direct contact between water and uranium, which would have imparted radioactivity to the water, the uranium cylinders, or slugs, were canned in aluminum. The problem was that excellent heat transfer between uranium and aluminum was necessary, or hot spots would develop and rupture the aluminum jackets. After much difficulty, a technique was perfected for injecting molten aluminum into the can as it was being filled with a uranium cylinder, and satisfactory heat transfer characteristics were thereby obtained.

Another problem was encountered at Hanford, even more ominous because it was unanticipated and threatened to thwart this entire path to the bomb. When the first reactor was completed and received its loading of uranium fuel, all went well for a few hours and then the plant proceeded to shut itself down. Several hours later, as if rested, the reactor began to perform properly again, only to repeat this puzzling behavior. The interpretation was that the U-235 in the fuel (which consisted of natural uranium) was producing numerous fission fragments, since it can split in a variety of ways, and one of these radioactive fragments, with a half life of several hours, was absorbing enough neutrons to interfere with the chain reaction. Since no more of this material was formed when the reactor was shut down, and since it decayed into a less sinister daughter product, when enough of it was transformed the reactor would start up again. It was at this point that the scientists came to appreciate the practical skills of the chemical engineers, who had insisted that the reactors be designed with a degree of flexibility.[15] In this case, a great many more fuel channels were built into the sides of the reactors, far more than the scientists had calculated would be needed to sustain a chain reaction. Now these holes were filled with uranium cylinders, and the reactor as a whole was able to function at a continuous level, producing neptunium and then plutonium through neutron capture by U-238, even if various parts within it fluctuated with the rise and fall of that unwanted fission product.

With fissionable U-235 and plutonium scheduled for production at Oak Ridge and Hanford, respectively, the next step was to plan a facility where these priceless metals would be fashioned into weapons. It was at first believed that only a small laboratory would be required, and consideration of water, power, and railroad connections gave way to isolation and secrecy. Los Alamos, New Mexico, was chosen, under the circumstances described within this book, but the proposed small facility of theoretical physicists, some experimentalists, and a few machinists, under J. Robert Oppenheimer's direction, soon expanded into a major installation of several thousand people.

Ballistics being a highly developed subject, there seemed little novelty about the design of a gun barrel in which one sub-critical piece of fissionable metal would be fired at another, the two joined producing a super-critical mass which would yield a nuclear explosion. Reliability tests for the components were, of course, necessary but this design gave no major troubles for U-235, and so confident were the scientists that it would function as planned that no test firing was conducted before the weapon was dropped on Hiroshima on 6 August 1945.

This technique proved hazardous for Pu-239, however. Another isotope of this element was discovered which fissioned spontaneously by neutrons produced by cosmic rays. To avoid predetonation in a gun barrel, the pieces might be fired at each other at the limits of ballistic velocities, but this was too uncertain a solution. Even less satisfactory would have been the separation of this offending isotope, for that would have removed the main reason for going ahead with plutonium manufacture – it could be chemically separated from uranium, and huge isotope separation facilities were not thought to be required. The solution perfected at Los Alamos was that of implosion. A sub-critical sphere of plutonium was surrounded by shaped charges of chemical high explosive. When this outer covering was detonated all over at the same instant, the plutonium core was squeezed into a tiny, super-dense super-critical mass in which neutrons did not have far to travel before striking fissionable nuclei. Uncertainty about the workability of this technique was so pervasive that a test firing was conducted near Alamogordo, New Mexico, on 16 July 1945. This was the famous Trinity Test[16] that startled those witnessing it by the enormity of the explosion. An implosion bomb was dropped on Nagasaki on 9 August 1945.

The lectures presented in this volume are based on this brief outline of wartime activities. But they are not intended to provide a thorough coverage of events at Los Alamos. Rather, they are personal recollections; they fill in much detail of only certain aspects of this history. A few points stand out. World War II was a 'popular' war, meaning that there was little dissent about the propriety of fighting against the Axis powers. This being so, and with the quite realistic assessment that Germany was capable of building (and using) nuclear weapons, there were virtually no refusals to join the Manhattan Project on moral grounds. With the war's end approaching, with Germany clearly defeated, and with most of their work accomplished, some visionaries, such as Leo Szilard and James Franck at the Chicago laboratory, urged that the weapons not be employed, or used only under certain conditions.[17] But those at Los Alamos had no time for such reflection. The frantic pace of their activity continued through the Japanese destruction.[18]

This book is neither an apology for nuclear weapons nor a glorification of them. It is an attempt to provide firsthand insights of a highly unusual period. Science and government were joined as never before, and the association has persisted and expanded in the postwar years. The wartime scientists recognized the change from their prior academic pursuits — from basic to applied research, from individual to organized efforts, from penury to prosperity in the laboratory budget — but few thought of it as a permanent alteration or questioned its need. Subsequently, many have reflected upon the role of nuclear weapons in our society and the dependence of science upon governmental support, but it was difficult to look for future implications when working overtime on problems of the present. Yet, despite their varied outlook today, it seems that the Los Alamos experience was the high point in most of their lives.

Why it was so emerges from these lectures. Beneath the descriptions of social events, housing problems, schooling for children, secrecy, etc., we see clearly the intensity of the period. Indeed, it would be hard to remain unaffected while working amidst a remarkable array of intellectual talent, on a weapon of enormous destructive capability, which might help to end the greatest war in history.

In such an environment it was inevitable that technical and personal problems would arise. Unlike most earlier accounts, some of these difficulties are explicitly discussed here — not just the familiar sniping at the Manhattan Project's military head, General Groves, but laboratory director J. Robert Oppenheimer's unwillingness to aid junior personnel caught bending security regulations a bit too much in Santa Fe bars; implosion inventor Seth Neddermeyer's reluctance to see his concept pursued on a large, crash-program scale; differences of style between Admiral Parsons and George Kistiakowsky; and the nadir of morale when the laboratory's future was in doubt, in the closing months of 1945.

These lectures also reveal some workings of the scientific community. Salaries were generally low and, being based on prior earnings, were not always in proportion to responsibility at Los Alamos. Another cause for complaint, one might expect, was the secret nature of the work, which prevented the normal process of publication and recognition operative in science. Yet, the reward system seems to have functioned after all, because the peers one cared about also were at Los Alamos and in a position to know of the contributions. It was as if the prewar international lines of communication, now called 'invisible colleges,' were condensed on the mesa for the duration. And it is worth noting that however much the Army was willing to build this

secret, isolated laboratory, it was the scientists themselves who recognized
its need and urged its construction. The *esprit de corps* they developed there
played no small role in their success.

Another aspect of life in a restricted environment centers on coping with
minor irritations and seeking diversionary entertainment. Whether planned
that way or not, General Groves, by dint of his personality, was an ideal
target for gripes about the bureaucracy. Food rationing, housing allocation,
mail censorship, and secrecy precautions provided focal points that served to
unite the community. This cohesion was built also by more pleasant activities, such as square dancing, theater productions, interaction with local
Indians, hiking, and skiing. With most of the professional community coming
from academic backgrounds, they were able quickly to adapt to each other
somewhat faster than to their new 'campus.'

Once there, and having solved a specific initial problem, the scientists
knew too much to be sent back to 'civilization.' Los Alamos was to be their
home for the war's duration, and so other tasks were placed before them.
That they succeeded so well when working on unfamiliar subjects speaks
highly for the breadth of their schooling. That they succeeded so rapidly that
the determining timetable of accomplishment was really the rate of production of fissionable material at Oak Ridge and Hanford speaks well for their
flexibility.

Most of the scientists there were physicists, since the fission reaction employed in both the plutonium and uranium-235 bombs fell within the scope
of nuclear physics. The list of speakers reflects this bias, but nevertheless
includes chemists, who were also vital to the project's success, wives of scientists, who provided unskilled labor at Los Alamos and furnish an important
perspective to these lectures, and the U.S. Army officer responsible for
choosing the laboratory's site. They give us not a technical or administrative
account of the New Mexico achievements, but rather personal vignettes of
the period 1943–1945. This being so we can accept different versions of the
same event, different impressions of individuals, and even the variant spellings
of Oppenheimer's nickname: Oppy and Oppie.

LAWRENCE BADASH

NOTES

[1] *New York Herald Tribune*, 12 Sept. 1933.
[2] Millikan, 'Alleged sins of science', *Scribner's Magazine* 87 (Feb. 1930), 119–20.
[3] *Pittsburgh Post-Gazette*, 29 Dec. 1934.

[4] For reproductions of the original papers, plus commentary, see Hans G. Graetzer and David L. Anderson (eds.), *The Discovery of Nuclear Fission* (New York: Van Nostrand Reinhold, 1971).

[5] For a survey of much of this literature, see Louis A. Turner, 'Nuclear fission', *Reviews of Modern Physics* 12 (Jan. 1940), 1–29.

[6] Charles Weiner, 'A new site for the seminar: the refugees and American physics in the thirties', *Perspectives in American History* 2 (1968), 190–234. Laura Fermi, *Illustrious Immigrants: The Intellectual Migration From Europe, 1930–1941* (Chicago: University Press, 1968).

[7] Daniel Kevles, *The Physicists* (New York: Knopf, 1978).

[8] Laura Fermi, *Atoms in the Family: My Life With Enrico Fermi* (Chicago: University Press, 1954).

[9] Leo Szilard, 'Reminiscences', *Perspectives in American History* 2 (1968), 94–151.

[10] Alan Beyerchen, *Scientists Under Hitler* (New Haven: Yale University Press, 1977).

[11] Samuel Goudsmit, *Alsos* (New York: Henry Schuman, 1947). David Irving, *The Virus House: Germany's Atomic Research and Allied Counter-measures* (London: Kimber, 1967).

[12] For the best detailed historical survey of the entire Manhattan Project, see Richard G. Hewlett and Oscar E. Anderson, *The New World, 1939/1946* (University Park: Pennsylvania State University Press, 1962). The official report released immediately after the war is Henry DeWolf Smyth, *Atomic Energy for Military Purposes* (Princeton: University Press, 1945). For a broader, but controversial, picture, see Robert Jungk, *Brighter Than a Thousand Suns* (London: Gollancz and Hart-Davis, 1958).

[13] Margaret Gowing, *Britain and Atomic Energy, 1939–1945* (London: Macmillan, 1964).

[14] A. H. Compton, *Atomic Quest* (London: Oxford University Press, 1956).

[15] For a discussion of the Manhattan Project that emphasizes the engineering, rather than the scientific, accomplishment, see Stephane Groueff, *Manhattan Project* (Boston: Little, Brown, 1967).

[16] Lansing Lamont, *Day of Trinity* (New York: Atheneum, 1965).

[17] The famous Franck Report, produced by a group of the Chicago scientists, may be found as an appendix in Jungk (note 12). Three interesting collections of articles about the history, politics, necessity, wisdom, and morality of the decision to use nuclear weapons against Japan are: Edwin Fogelman (ed.), *Hiroshima: The Decision to Use the A-Bomb* (New York: Scribner's, 1964); Paul R. Baker (ed.), *The Atomic Bomb: The Great Decision* (New York: Holt, Rinehart and Winston, 1968); Barton J. Bernstein (ed.), *The Atomic Bomb: The Critical Issues* (Boston: Little, Brown, 1976).

[18] From the large literature about the effects of nuclear weapons, the following may be cited: United States Strategic Bombing Survey, *The Effects of Atomic Bombs on Hiroshima and Nagasaki* (Washington, D.C.: Government Printing Office, 1946); John Hersey, *Hiroshima* (Harmondsworth, Middlesex: Penguin Books, 1946); Robert Jay Lifton, *Death in Life: Survivors of Hiroshima* (New York: Random House, 1967). For the reaction to the use of nuclear weapons, see Erwin N. Hiebert, *The Impact of Atomic Energy: A History of Responses by Governments, Scientists, and Religious Groups* (Newton, Kansas: Faith and Life Press, 1961); Alice Kimball Smith, *A Peril and a Hope: The Scientists' Movement in America: 1945–47* (Chicago: University Press, 1965); Morton Grodzins and Eugene Rabinowitch (eds.), *The Atomic Age: Scientists in National and World Affairs* (New York: Basic Books, 1963).

JOHN H. DUDLEY

RANCH SCHOOL TO SECRET CITY

As we go through life, we find that there are different points of view on most matters. For example, if there are two eye witnesses to the same event there is great probability that they will have different stories; both are quite honest. A number of years ago when I was at the Armed Forces Staff College I prepared a paper on one military operation. As sources I had eye witness accounts of five different people. The only similarities between them were the date and place. Beyond that you would think they were writing about entirely different events.

This illustration has bearing upon the procedure for the selection of Los Alamos as the secret laboratory site. I shall not describe only one procedure; I am going to tell you four different accounts of the selection, each of which can be proven. You may then select your favorite.

The military services in this country became interested in atomic power about 1939, and at that time we put some money into it. The scientists tend to sneer at the amount but for the time it was a large sum, or a large proportion of the money we had for scientific work. Interestingly enough, the concern was with propulsion. Propulsion is important, very important in military operations. Mobility stems from the power you have available to move around. If you do not have mobility, you get nowhere, no matter what kind of weapons you have. Going back several centuries, history records the invention of gunpowder. Gunpowder is credited with having a major effect on warfare. But some centuries earlier there was another invention. This was the lowly stirrup, that little thing that hangs on the side of the saddle when you ride a horse. The invention of that had a dramatic effect upon the capabilities of the people in those days to use horses. Many believe that it probably had a more dramatic effect on military operations and the history of nations of that period than the invention of gunpowder had in its own time.

The scientists in 1940—41 were afraid that Hitler might build an atomic bomb. As a result, they felt that they must try to make a bomb first, and took various steps, such as bringing their fears to the government and introducing secrecy in uranium work.

By mid-1942, President Franklin Roosevelt issued instructions to the Army Corps of Engineers to construct the factories needed to manufacture the

1

L. Badash, J. O. Hirschfelder and H. P. Broida (eds.), Reminiscences of Los Alamos 1943–1945, 1–11.
Copyright © 1980 by D. Reidel Publishing Company.

fissionable materials and then to manufacture the bomb. The Corps of Engineers probably was picked because of its experience in constructing factories and moving rapidly. The latter ability could be accomplished most effectively through its extensive organization. The system was for the Chief of Engineers to pick a key man from some place around the country, assign him the job, and then help him select other key people, all to be made immediately available. In other words they did not send out want ads to look for the initial team. A first team was picked composed of people who had already demonstrated their abilities and capacity to work together.

In this case the Chief of Engineers picked Colonel J. C. Marshall to head the new organization. Marshall was the District Engineer in Syracuse, New York, which district had been organized a few years earlier by this first team rush type of procedure. He had had extensive experience in Washington and had constructed a number of complex factories and other works. A few hours later, in Washington, he found out what the job assignment was and also that there was no available office space in the capital. But since he was going to construct factories, New York City, with its industrial headquarters, was probably a better location for him. When he looked at the sort of plants he was to build he found that the basic process research had not been completed. Usually in developing a new product the sequence is: basic research, applied research, construction of a table model, construction of a pilot plant, engineering for the real factory and then construction of the factory. This is a long chain of steps, and only the first was in progress. He tried to persuade the people involved to speed up, to get the early steps done.

At the same time the scientists desired that the head office on atomic matters be in Washington, for ease of contact with the government. They also felt that a lieutenant general should head this work and were not quite satisfied with a colonel. Lieutenant generals at that time were very few, and still are scarce; none could be made available. Finally they settled on Leslie Groves who was a brigadier general. The addition of Groves made Marshall redundant and after somewhat over a year he was able to arrange to move on to other duties. Actually, for a while it looked as though he was stuck in the Manhattan District because the rule, when he started out, was that if you were assigned to this project you were prohibited from leaving it for other work until the war was over. However, with the passage of time there were a few who were able to leave. Also the objective in manning the job was to use only a very few active regular army officers: only two for the entire district. The number did go up to five but this was temporary. The majority of the officers that the scientists ran into were retired officers recalled to active duty, reserve officers and temporary officers.

My first connection with this work was in the very pleasant fall weather of October, 1942. I was borrowed from my district to travel around the West to locate a site for a mysterious installation. At that time, in fact all the way through, I was not supposed to know its purpose. I was simply given a set of criteria to follow. There was no particular advantage to my knowing, and actually it was to my advantage not to know, the purpose of the installation. Officially, I was able to maintain this fiction of ignorance throughout my association with the Manhattan Project. However, as soon as I started talking to scientists, say within a matter of a half hour or forty-five minutes, I knew what they were doing quite thoroughly. I tried to turn them off, but within two minutes they would again be leaking information.

The principal site criteria were:

Population 265. This number apparently was based upon Oppenheimer's idea that 6 scientists, assisted by some engineers, technicians, and draftsmen, could do the job quite rapidly and effectively. These people would have their families with them. In addition there would be a need for some plumbers, electricians, guards, storekeepers, school teachers for the children, and people like that. These would bring the total up to 265. After talking to Drs. Ernest Lawrence and Robert Oppenheimer in Berkeley in the course of my travels, I raised the number to 450. Later, at the end of November, we arbitrarily raised the figure to 600 for planning purposes. That was in order to arrange for sufficient water supply, housing accommodations, and things of that nature.

Inland over 200 miles. The site had to be at least two hundred miles from the Pacific coast and from any international boundary, while staying west of the Mississippi. That ruled out just about all of California and the southern portions of Arizona and New Mexico. The requirement that the winter climate be pleasant ruled out Idaho, Montana and some other places.

Some facilities. Some existing facilities were needed so that a few workers could move in right away. The idea was that the 6 scientists would move in and start a think-tank operation immediately. Some space was needed for that. In looking for places with some facilities you naturally consider resort hotels and dude ranches. However the Air Force and the Navy had been through that area some months before picking up the best ones for use as rest centers. There wasn't much left over.

Isolated, accessible. These terms actually were not contradictory. Accessible meant that a good road connection to a railroad was required so that supplies could be hauled in easily. It had to be fairly close to a commercial airport so that people from Washington, Berkeley, New York could fly in and out for conferences. Isolated meant that a rural atmosphere was desired, with sparse population — a type of population where strangers would be quickly noticed.

Natural bowl, hills close. It had to be in a natural bowl with hills or mountains close in. The idea was that we would put the fence up on top of the ridge and patrol the fence, keeping all strangers out so that they couldn't even peek over the hill top to observe what was going on, or even count how many people were there. To a few of us the bowl had some elements of a safety provision because an explosion would wipe out a few hundred people instead of thousands.

There were other criteria but the above were the most significant.

Toward the end of October 1942 I started out on my search. My plan was to visit the various Corps of Engineer offices in the southwest. I carried with me a letter from the Chief of Engineers that said "give this man all the help he needs". This Corps system is very helpful: you can go into a strange office, meet people you have never seen before, and you do not have to negotiate a contract or a purchase order for them to work with you. You can get someone who knows the region thoroughly, who is accustomed to talking in the same terms you use, and who does not insist on asking a lot of unnecessary questions. Elsewhere you find people who won't move unless you answer delicate questions about what you are doing.

I traveled by air, rail and auto. Perhaps a thousand miles were covered on two-lane roads — one lane for the left wheels and one lane for the right wheels. When the going got tough, I switched to a jeep, and when it got even tougher, I rode a horse.

I toured through parts of California, Nevada, Utah, Arizona and New Mexico. The best place I found that fitted the criteria was Oak City, Utah. It is a delightful little oasis in south central Utah. The railroad was only 16 miles away, over a nice, easy road. The airport was not too distant. The water supply was good. It was surrounded by hills, and beyond there was mostly desert. However, I noticed one thing: if we took over this area we would evict several dozen families and we would also take a large amount of farm acreage out of production. So I recommended to Colonel Marshall that we skip over Oak City to avoid these adverse effects.

Instead, I suggested the second choice place, Jemez Springs, New Mexico.

The access road grade was fairly easy so that heavy hauling would be no problem. There were some existing facilities, including a resort hotel, some empty housing, and other empty buildings. There were very few people who would have to be evicted. Colonel Marshall ok'd this area for a site study, which meant that we would set down the details of what the land was like, who was there, and things like that.

We arranged for Drs. Oppenheimer and Ed McMillan to meet me at Jemez Springs to inspect it. One of the things I liked about the place was the small ridge between the technical area where the 6 scientists would be working and where the families would live. If the place blew up only the 6 scientists would be involved and not all the families. Oppenheimer considered the idea of no value. He took one look at Jemez Springs and proceeded to change the criteria. He said, "no hills around", while I had been hunting for hills. As for the access road, he just needed it good enough to haul two heavy howitzers up the road. Now, I could have taken a heavy howitzer up a dirt track. So, the problem of access was practically wiped out.

On meeting us at Jemez Springs, Groves was unhappy that we did not have a site picked. So he asked Oppenheimer if there was something else around that had prospects. Oppy proposed Los Alamos as though it was a brand new idea. As I had been at Los Alamos twice and knew the roads (or trails) from Jemez Springs to Los Alamos, we drove directly there. On looking it over Groves was not too pleased with it. But, having business down in Albuquerque he left us while McMillan, Oppenheimer and I had dinner someplace. I think it was in Santa Fe. Later that evening, on seeing Groves again, I told him the general tenor of our dinner conversation. This was that the criteria would be continuously modified until only Los Alamos fit the criteria. So I said, "Let's pick Los Alamos and get on with our other work." Groves approved and phoned immediately to his secretary back in Washington, Mrs. O'Leary, telling her to start the acquisition procedures the next day.

There are, however, other stories of how the place was selected. The second one concerns P. C. Keith, who was a senior manager in the Kellex Corporation, doing much work for the Manhattan District, and a member of OSRD (Office of Scientific Research and Development in Washington). He had two sons at the Los Alamos Ranch School, knew the school's financial problems and their difficulty in getting suitable instructors. They were running into major problems in staffing the school. Keith suggested Los Alamos to Colonel Marshall. Colonel Marshall made a visit to it in mid-1942, picked the site and that was it. Colonel Marshall, who later become a general, denies that he was ever there until 1943. But we have eye witness accounts that say he was there!

The third story concerns Colonel L. S. Hitchcock who had been the headmaster at the Ranch School but was called back to active duty. Stationed in Washington, he knew Groves. In the Spring of 1942 he told Colonel Groves (then a colonel) about the place. Groves went out to Albuquerque, borrowed a car, drove by himself up into the wilds, stuck the car in rough ground, walked out, and went back east. However at that time Groves was not connected with the project in any way. He was much too busy with other things to take a little side trip. It is not his nature to drive a government vehicle and he says that he never saw the place before November 1942.

The fourth story. In the course of a seminar at Princeton in the early 1950's Dr. Oppenheimer made a statement about the selection of the place. When he was asked to head up the laboratory, he determined that the location would be Los Alamos, or he would say no. Los Alamos — Oppenheimer, or not Los Alamos — somebody else.

You now have four stories from which to pick.

We proceeded with the acquisition of our installation but we couldn't get full access for two months as the school had to finish out its year. However there were a lot of preparations needed. Oppenheimer had to get together a technical staff of which you undoubtedly have heard. Considerable construction had to be planned and started; there were many problems in the construction. And we had to get a post commander.

The criteria for the post commander indicated that he have ability and experience in management, that he get along with scientists, and that he had to be a bachelor (although a widower would be acceptable). Our best prospect owned and played an organ. Although we tried to entice him by finding a place for his organ, he slipped away. In spite of my being married, I was offered the job twice and turned it down. This probably lengthened my life somewhat. Finally they picked Colonel Harmon from another important job and moved him in.

We set up an office in Santa Fe on the 4th of January, 1943. A few days later we set up an office for Oppenheimer in another building. The separation was necessary as the office buildings in Santa Fe at that time were quite small; there wasn't space for the two together. Later I came to realize that actually it was probably a good idea that we were separate.

The initial construction contract looked like 300 thousand dollars. Within a year that first contract had gone to 7½ million, a 25-fold increase. And it was only the first contract; there were many others because there were major difficulties in the construction and major expansions.

We had trouble getting the support personnel to help the post commander.

The types we needed were in short supply. We solved some of that by having more military personnel there than would be our normal habit, since the military service was the only place where we could find many of them.

The special Military Police Company arrived in mid-April. I ran into considerable obstacles in Washington in setting it up because a number of desk-warming majors there gave interference. I had a letter from their boss saying that they were to give me anything that I wanted but they repeatedly ignored the letter and would say "no", they couldn't do it. Then I would have to show them how they could do it, and threaten to take the problem back to their boss. I finally got what I was after in this special unit. This MP Company, interestingly, brought along a lieutenant whom I had known years previously as a sergeant in one of my outfits. It was a happy reunion.

A WAC detachment (Women's Army Corp) arrived toward the end of April. They had a very strange trip from their base in Iowa to Los Alamos.

Briefly, they started out with secret orders to an Army base in Oklahoma. On arrival there they were greeted, given a place to sleep for the night, and the next day given other secret orders to a place in Texas. On arrival they received another change of secret orders to report to a certain room in an office building in Santa Fe, New Mexico. Then their train, headed for Albuquerque (passenger trains did not run to Santa Fe), was blocked about 15 miles south of there for perhaps 18 hours.

Expecting to meet them in Albuquerque, I found that they were not going to get in and drove on down to pick them up at a little town to the south. When I introduced myself to the officer in charge she had reason to be suspicious. Here was a strange lieutenant colonel; he was driving himself (and you know in the Army the officers do not drive) and he was driving a civilian government car. The second vehicle, also civilian, had a civilian driver. But they decided that maybe we were honest and came along with us. We had early breakfast in Albuquerque, and passed through Santa Fe just about at the first streak of daylight — it was a small town then with nice sleepy streets.

At dawn we were driving up wild canyon country and passed through a crevice in the rocks, very similar to those described in some of Zane Grey's stories. Then an extensive community appeared in view. On arrival at Los Alamos, shortly after dawn, the WACs moved into barracks, which, to their surprise, were deluxe compared to the standards to which they had become accustomed.

A high level meeting took place at Los Alamos early in March. Although I was not included in it, my impression was that the group decided that the place was being improperly manned and improperly equipped. A tremendous program was started to increase the size, add more buildings, more offices and shops, more housing and so forth. I got this program started and then was able to make use of the fiction that I did not know the site's mission. Early in May I left to take command of an engineer regiment and ended my connections with Los Alamos.

By way of an epilogue, my unit and I moved through Papua, Admiralty Islands, Netherlands New Guinea, Spice Islands, and the Philippines. In July and August 1945, we were in the northern end of Luzon, an island nearly the size of California. In early August I made a trip to Sixth Army Headquarters near Manila to help with the detailed planning of our next operation. These were the plans for the landing on Kyushu, Japan. To my surprise, my unit was scheduled to go in three weeks after the initial landing, on the turnaround shipping. As we had been through five operations with a reputation of regularly going in on the first landing, I was quite startled. In looking over the details of the plan I discovered that the second group had essentially the same mission as the first group. The expectation was that the first group would largely be wiped out, and the second group would carry on. Considering the other landings that would be necessary, especially the one in central Honshu, apparently the expectation was that about 200 to 300 thousand Americans would be killed in the operation. The Japanese are tenacious and courageous; maybe 10 million of them would be killed. Bad things on bad things. A few days later: 6 August – Hiroshima, 9 August – Nagasaki, 14 August – peace. A vast sigh of relief went through the area.

DISCUSSION

Question: I've sat through a number of such lectures and heard about how recalcitrant the military was. Being retired military myself I had to bite my tongue. Perhaps you could give us the other point of view.

Dudley: I would say that my relations with the scientists were quite good. Whenever you have a large operation going on, things moving ahead, there are disagreements. We had disagreements with scientists. Remember, many of these scientists had relatively little experience. They knew their nuclear physics or they knew their chemistry but they did not know what life was like. So you have problems. But as a whole our troubles were tempests in a teapot. I had as much trouble with those three majors in Washington. I had trouble with Groves. One time he wanted me to do something which was illegal. I told him it was illegal and I wasn't going to do it. He looked surprised, said OK and changed his orders. When you have things moving ahead, you do get disagreements and the scientists, as I say, were unusual people. Oppenheimer himself had a broad experience and I could work things out with him usually quite well.

Now I know that there were some special complaints, like housing. Why did we put those horrible houses there? An Act of Congress about early 1942 established a civilian housing agency which set up standards for housing in the United States to be built during the war years. They specified what would go on the inside. For instance they specified 'showers, no bathtubs'. The manufacturers of bathtubs in the United States had ceased manufacturing bathtubs about 1942. So even if you went to try to get them, they were hard to find. The orders included putting in coal stoves. Some of us have had experience in cooking on a coal stove; not very many. It can be done! Very fine pies, cakes, roasts and things like that have come out of coal stoves. We did find, up one of the side valleys of Santa Fe, an inventor who helped us with the stoves. Santa Fe was a very delightful place. Many people had gone there because of the calmness and peace around there. This individual had been the chief engineer on the first round-the-world aerial flight. He had retired in that area and was doing a little inventing. We found that he had a device burning diesel fuel to put inside the coal stove. We could match the criteria set by the civilian agency in Washington while putting in these inserts. Incidentally that insert later came of value to me in Japan in 1945. We got hundreds of them and they were life savers in keeping us warm. So I was glad that we had found them. Along about January-February of 1943 the housing criteria were outside our hands. Groves tried to get the criteria modified. He was successful in many matters but not in the housing problem.

Question: One of your criteria was a pleasant winter climate. Surely that wasn't the scientists' plan?
Dudley: If you must recruit people who can go voluntarily to other places, you don't want to ask them to move into a bad winter climate. Los Alamos, as you know, does get some snow, but most of the year it is a very healthy and pleasant place.

Question: This has no connection with the natural work outside all year?
Dudley: I don't remember any connection with having to work outside all year. It may have been.

Question: What were some of your conventional civil engineering problems?
Dudley: We had to haul a large diesel electric generator up a poor road. That was no particular problem. As this engine block was coming in on a trailer,

the truck driver had instructions to stop at the little town of Española, to phone ahead for instructions and for assistance. When he got there, he looked down the road and saw that it looked easy. So he just drove on. After some turns the road became worse until he dumped the engine block off his trailer in the middle of the road. Picking up a piece like that is not too bad a job except that in dropping it, he cracked the block. It was an unique item, very, very difficult to replace in those days. The contractor was afraid to pick it up until he repaired the block. So he built a house over the block, constructed a detour around it, and proceeded with a patching operation that took about a month. Then he tore the house down and moved the engine block on in.

For a population of 265 we did not need to do much excavation, but when we started getting up into the thousands, we had miles and miles of rock excavation for utility trenches. Rock excavation is extremely expensive. Beyond that the main job was to push straight through. Personnel matters were another problem. However the type of construction was straightforward, nothing unique. Installing the cyclotron that the scientists swiped from Harvard was only a minor problem. With a few electricians accustomed to constructing or installing electrical equipment it was put together readily.

Question: Did you install any sophisticated intrusion preventive barriers?
Dudley: We didn't have anything like those snoopers that you could leave out in the woods to tell when somebody came by. Nothing like that. We had to rely upon observation and that the very sparse population around us would tell us if there were strangers in the area. We did have a possible problem with ham radio operators who were not permitted to take their transmitters into the site. But there was enough equipment in there to construct two or three dozen transmitters, which would have been no problem to many there. So we did have a radio intercept system set up so that we could have intercepted radio messages going out.

We only had two telephones. Oppenheimer's idea was one telephone for himself, one for the post commander, and any volume business would go out over a teletype. Oppenheimer would control what went out on his phone and the post commander what went out on his so that the telephone was no problem. Later we added on a lot of circuits, but their use was after my time.

Question: Was there a proposal to put the scientists in uniform?
Dudley: Probably from Oppenheimer I heard that it had been considered and had been dropped by the wayside. Some people said that because it was a

military installation they had to be in uniform. But I do not think that was actually of any importance because the Corps of Engineers had had a century and a half of experience of working with mixed groups of civilians and military. The two together made no problem for the management people. I think there would have been a problem for Oppenheimer in recruiting because a portion of the scientists would refuse to participate if they had to get into uniform. In connection with morale I think it was better not to have them in uniform. As you may realize by now, in academe there is much rank consciousness. If you mix assistant professors, associate professors, majors, and lieutenant colonels all together you are bound to have all sorts of troubles. I have an idea that these men, coming in fresh without any military training at all, would be unwilling to accept the additional responsibilities that go with rank. Rank is not all a smooth bed of roses. You acquire responsibilities with it. I don't think many of them would have been willing to accept the responsibilities. I think that it was better that the place stayed on a civilian basis, although some, like Admiral Parsons, arrived later. I do not know what other people have said about the arrangement but I expect that it worked all right.

Question: What was your reaction when, unofficially, you learned what the project was about?
Dudley: My reaction was surprise that they were that far along. And it turned out that I was entitled to be surprised. I do not recall how I knew about nuclear fission. Undoubtedly in my undergraduate days I had a little bit of quantum mechanics of an elementary type. I doubt that I got any at M.I.T. At an Army school I attended in 1939–40 I may have picked up something. But the subject matter was no surprise to me.

Question: You mentioned your involvement in the Pacific after leaving Los Alamos. What were the prospects in the minds of the military as far as the potential part that Russia might play in Japan had the bombs not been used?
Dudley: I didn't think about it and I do not remember other people doing so. We were very busy with our immediate responsibilities. It didn't enter our picture.

EDWIN M. McMILLAN

EARLY DAYS AT LOS ALAMOS

I must emphasize the difficulty of establishing facts at this late date, even of important things. During the Manhattan Project, of course, security was impressed upon everyone. Very few people kept any notes; we didn't keep diaries or little black books with all the records of where we'd been and what we'd done, and I found that other people had the same difficulty remembering. But trying to establish detailed facts turned out to be quite an interesting thing. There is a pre-history to Los Alamos as well as a history. I shall describe what occurred between the specific dates of September 14, 1942 and September 17, 1943, almost exactly a year later.

On the first date I was at the U.S. Navy Radio and Sound Laboratory in San Diego, working on sonar, when I got a telegram from Arthur Compton in Chicago, asking me to come to a meeting one week later with Robert Oppenheimer, John Manley, Enrico Fermi, Ernest Lawrence, and others. Between my receipt of that telegram and my attendance of the meeting, a very important thing happened. On September 17, General Groves joined the project. He was asked to take over the Manhattan Engineer District, which already existed, he consented to do so, and was present at the meeting in Chicago. At this meeting I decided I would join the project. I had already been approached by Lawrence in Berkeley, who wanted me to come and help him in his isotope separation work, and I had begged off on that for at least a few months until I finished the project I had going at San Diego. But learning about the other ramifications of the project and the starting of a new laboratory — that was very attractive to me. Also, Compton wanted me to come to Chicago and work with him, but the new laboratory was the magnet that drew me, something somewhat adventurous, and I decided to join the project. Shortly after that, I moved from San Diego to Berkeley and my wife moved up with her parents in San Marino, and I spent the next few months in Berkeley working with Oppenheimer and Serber and a number of people who came and went during this period.

My office in Le Conte Hall at the University of California at Berkeley was the organizing center for the laboratory, and in that office we made plans, mostly preliminary plans, but some things had to be done in more detail. During this same time, John Manley, working out of Chicago, was also involved in

13

L. Badash, J. O. Hirschfelder and H. P. Broida (eds.), Reminiscences of Los Alamos
1943–1945, 13–19.

the planning, and I find some correspondence between us. We had to decide what equipment was needed, think about personnel, and also come up with some idea of what laboratory buildings were needed. Colonel Dudley, who was attached to the Army Engineers, and General Groves, who was by then already well established in the project, came to Berkeley and met with us; they were talking principally about possible sites and the buildings that would go there, and I remember looking at many architect's plans showing what they suggested as to how things were to be arranged, but as of that time, (November 1st and 4th were the dates of these two visits by Dudley and Groves) the site had not been chosen. It had to be far from the borders of the United States for security reasons, and many places were looked at. This site search ended when Colonel Dudley had decided, on the basis of the criteria and of what he had seen, that the best site was at Jemez Springs, New Mexico.

Oppenheimer and I arranged to go there. We were supposed to get horses, Dudley was to meet us, and we were to ride the boundary. The idea was to see whether the security was adequate, where the fences were to be put and all that kind of thing, and also to look at existing buildings and the general adequacy of the site. The date of this visit took a lot of research to find out because nobody remembered when it was. Finally it got tied down. General Groves found his diary — he did keep a diary and located it. Colonel Dudley had turned over all of his records to the War Department and they'd been put in the archives and then at some time, in what is called a 'reduction in volume of records', they were all destroyed. His personal logs of where he had been were all destroyed. That was a historical tragedy. Then he discovered that he had some boxes in which he'd kept a few things out, like travel vouchers and stuff like that, and from those boxes he was able to find dates of where he'd been at various times, and that confirmed this date, which was November 16, 1942.

Jemez Springs is in a canyon that runs westward out of the Jemez Mountains and it's a deep canyon with a flat bottom. It's got a river and a road and a little bit of land, sort of a one-dimensional site. It didn't seem to me that there was room there, and I imagined a community all strung out like that; it seemed very awkward and I expressed considerable reservations about this site. Of course Dudley was supporting it very strongly because it was his choice and it was, I suppose, the only one that really met all the criteria exactly. For one thing we couldn't ride the boundaries; that was impossible because the fence would be way up on the top of the canyon wall. It would be several days work to get around, and I'm not even sure you could take a horse around there. We were arguing about this when General Groves showed up. This had been planned. He would come in sometime in the afternoon and receive our report. As soon as Groves saw the site he didn't like it; he said, "This will never do," and I never had a chance to make my fine argument about it. At that point Oppenheimer spoke up and said "if you go on up the canyon you come out on top of the mesa and there's a boys' school there which might be a usable site." So we all got into cars, we didn't ride horses, as some people have said, (that's too far) and we went up to Los Alamos Ranch School.

I remember arriving there and it was late in the afternoon. There was a slight snow falling; it was just a tiny, drizzly type of snow. It was cold and there were the boys and their masters out on the playing fields in shorts. I remarked that they really believed in hardening up the youth. As soon as Groves saw it, he said, in effect, "This is the place." In a letter to me, Groves wrote: "I think it perfectly correct to state that there were two decisions made on this trip. The first was turning down Jemez Springs. The second was the selection of Los Alamos. I had no intimation before turning down Jemez Springs that Los Alamos was a possible site." As soon as the site was chosen a report was authorized so they could make the official investigation and arrange for acquisition; this went through rather quickly and the site was acquired. I believe that the ranch school was not in great financial condition, anyhow, so there was no reluctance to sell. Then I went on to Washington with Oppenheimer. We had some discussions there, came back and four days later actually went into Los Alamos. The first time nobody went in, we just stood outside, as I remember, and looked at it. The second time it was Oppenheimer, Lawrence and myself, on November 20th, who were among the first to inspect the site. Of course Oppenheimer knew it already because, having a ranch in that area in the Sangre de Cristo Mountains, he had been over and seen the school before. I think Oppenheimer privately wanted that site anyhow but he didn't put it forward until this occasion; another story I've heard was that Los Alamos had been chosen all the time and the Jemez Springs visit was just an exercise, but Groves' statement that he had not heard of Los Alamos as a site before would put that one down. One objection to Los Alamos was that it didn't have mountains around it. But on the other hand it was on top of a mountain, or rather a mesa, which was just as good. There were other objections such as inadequate roads, water and so on but they were not fatal and they were overcome.

Later in 1942 and early in 1943 many things went on, not only the site and research planning, but ordering equipment. I remember writing out orders for machine tools to set up a large shop, something I really didn't know much about, but I had the advice of the shop man at Berkeley and somebody had to do it, so that when the shop was there the tools would be ordered. There's a long lead time on some of those things. Another thing I remember was recruiting. I was asked to go to Princeton where Robert Wilson had a group working on a method for separating isotopes. I talked to these people and recruited essentially the whole Wilson group. That must have been about 20 people or so, including Bob Wilson and Dick Feynman. Another thing I recall was a trip around the country to laboratories that had cyclotrons to choose one for the project. Los Alamos had to get equipment, nuclear equipment, and I went to several laboratories that had cyclotrons of the right size, and the one that was best suited, in the best condition, portable, or as close to portable as a cyclotron gets, was the Harvard cyclotron. On my recommendation it was taken, and for many years after that some people at Harvard were very angry with me. They would say, "You stole our cyclotron." But then it did very valuable work at Los Alamos and so I think it was worthwhile. Of course, in the meantime, construction was going on, housing, laboratories and so on. Oppenheimer and his immediate staff arrived on March 15, 1943, so from the middle of March perhaps 20 research people were there. I followed very shortly after that, Mrs. McMillan came on April 1, and from then on the laboratory grew rapidly.

I was connected in a small way with the early stages of the implosion program. There were two methods of bomb assembly that were worked on. One was called the gun method, which consists of simply firing two chunks of fissionable material together at high speed and then releasing some neutrons to start the chain reaction. The other was the implosion method, where a converging explosive wave pushes material in from all directions. This has the advantage of being faster, and there are other physical reasons why it is a better way. But it was by no means clear at that point that one could implode a material and have the implosion go in nice and smoothly so it all goes to one point rather than crumpling up in some complicated way which would increase the critical mass. The beginning of the implosion program was at a meeting at Los Alamos, which must have been in late March or early April of 1943, at which Seth Neddermeyer, who joined the project from Cal Tech, proposed the idea. As far as I know nobody else had thought about it; the gun method was *the* method. There was a lot of skepticism about it, whether it was practical at all, but Seth wanted to get on with the job and try it out. So without any particular official recognition from the laboratory he set up to do the early work on his own. He went to Bruceton, Pennsylvania, where the Bureau of Mines had an explosives research station, to learn something about explosions, and I went with him, as I was very interested in this idea. While there we saw George Kistiakowsky. He wasn't in the project at that point, but was connected with the Bruceton laboratory. I had known him before from Princeton. Later, of course, he took over the implosion program, but at that point it was just Seth and myself with a few helpers. The first cylindrical implosions were done at Bruceton. You take a piece of iron pipe, wrap the explosives around it, and ignite it at several points so that you get a converging wave and squash the cylinder in. That was the birth of the experimental work on implosion, long before experimental work on the gun method, a point which is sometimes obscured in the histories.

It was a very interesting trip, and when we came back Neddermeyer was able to get enough support to set up a little research station on what was called South Mesa, just south of the main technical area at Los Alamos, for doing implosion work, and I assisted him on that. Explosives had to be made into odd shapes so that we needed something that could be manipulated, and the things we used then were powdered TNT and plastic explosive, the same thing that terrorists like to use now. It looks just like putty, and you can mold it into any shape you want. One experience there made me feel so stupid that I never told anybody about it until years later. I was carrying powdered TNT in a wooden box about a foot and a half long with quite a lot of the stuff in it, with an open top. At that time I was a cigarette smoker. I was walking through the woods when I suddenly realized there was a lighted cigarette bobbing up and down in my mouth with that big box of powdered TNT underneath. I couldn't release a hand to take the cigarette out of my mouth so I had to put the box down (very carefully!) before I could get rid of the cigarette, and when I delivered my burden at the test station I told no one about what I had done. Those tests of course could not be very sophisticated; they were all in cylindrical form, as I remember, nothing as complicated as a sphere was done. They did show that you could take metal pipes and close them right in so that they became like solid bars, indicating that this was a practical method; of course you know that from later events.

I became associated with the gun program when Captain Parsons of the Navy was put into the project to take over that work and I became his deputy. I even shared an office with him for quite a while when we didn't have space. The first task of the gun group was to set up a test stand where experiments could be done. You have to have a gun emplacement, and a gun, and a sand butt, which is nothing but a huge box full of sand that you fire projectiles into so that you can find the pieces afterwards, and because there might be somebody else out there. The Anchor Ranch test range was set up;

it was probably over-designed but everybody was very worried at that time about guns blowing up, and there were all kinds of horror stories about the dire consequences of that. Anchor Ranch was one of the old ranches, not part of the Ranch School but in the vicinity, and the owner of the ranch was moved out and he'd left everything behind. It was a complete ranch with house, barn, equipment, everything including a flat area which would make a good test range, next to a small canyon so the control building could be down in the canyon with the gun on the flat above. After the gun was loaded you would go down into the control room, which had thick concrete walls, fire the gun from there, and observe what happened through a periscope. But more important, you would go out afterwards and look at the pieces, which would tell you how the projectile and whatever it hit had behaved in the impact. In gunnery they have something called a yaw card, which is just a piece of cardboard that is put out in front of a gun, and which the projectile goes through so fast that it cuts a hole just like the projection of the projectile. If it is going sideways it makes a rectangular hole (if it is flat-nosed) or a round hole if it's going forward, and if it's something complicated that breaks up, every piece makes its mark. There would be a yaw card, and a crew who would get out their shovels and dig down in all that sand to find the pieces so that they could be examined, and also various more sophisticated diagnostic devices. The first shot at Anchor Ranch was fired on September 17, 1943, which I have chosen to represent the transition from early to late history.

A couple of little stories remain in my memory. In running a laboratory in a remote site there are many things that you have to do in a different way than you do ordinarily. For instance, power. Here, if you want power, you just hook on to the local power company; a company crew comes and connects you to the line and you throw a switch. At Los Alamos there was no local power company, so some old diesel-run generators were acquired from an abandoned mine, I believe in Colorado, and were set up on the site. There were as I remember four of these generators; they were all hooked together to furnish power to the laboratory. They were run by one of those grizzled old timers who had been doing this kind of thing for many years; I think he came along with the generators, in fact. The plant ran, but with some interesting peculiarities. One of the things which you like to have for running scientific equipment is stable frequency and voltage in the power supply. There were complaints about fluctuations, so I had in my office two Esterline-Angus recorders, one recording frequency and one recording voltage continuously. The frequency was wandering back and forth too much. The power committee consisted of three people, John Williams, who was chairman, Bob Wilson and myself, and we went down to the power plant and conferred with this man. He said: "we have a master clock, you know, with two hands. One hand is run by a pendulum clock and the other is run by a synchronous clock connected to the power line, and you're supposed to keep those hands going around together. There are some little weights on the pendulum that can be changed and I keep changing the weights and I always keep these two hands together." The other story also involves frequency. The laboratory grew in size and the power plant became inadequate, so there was talk of running a line up from the Rio Grande valley where there existed a sufficient power supply. This man said: "that will never work, you'll lose so many cycles going up that hill that you'll never be able to synchronize them."

JOHN H. MANLEY

A NEW LABORATORY IS BORN

I have learned from my historian friends that I am classified as a "living source material." I have never quite figured out where in the hierarchy of honors being "source material" comes, but when one gets to my age, the adjective "living" is encouraging at least.

I hope to describe the background to my association with the Manhattan Project, before discussing Los Alamos. Out of this will come a picture of the small world of physics in the 1930's, where people of my generation knew most of the leaders and had intimate contact with many of them.

I think it's fair to say that the start on the road to Los Alamos was an appointment at Columbia University following my doctoral work at Ann Arbor. At Michigan I'd been interested in atomic physics, and I don't mean atomic bomb physics at all, but physics of the atom and not physics of the nucleus. But, on moving to Columbia, I did start to get interested in nuclear physics and nuclear physics was the basic physics for the bomb. At first when I went to Columbia I worked with I. I. Rabi, making measurements of particular nuclear properties with a molecular and atomic beam method for which Rabi later got the Nobel Prize. My own feeling, I remember at the time, was that I didn't care too much about New York City and that the particular research work Rabi engaged in was too specialized for my taste and not easy to take somewhere else if I were to move. Prompted by this feeling, I started looking around for other things that were going on at Columbia. I had gone there in 1934, and perhaps after a year and a half working with Rabi I shifted into neutron physics. Neutron physics at Columbia was a very active area. John Dunning had some people building a cyclotron in the basement of Pupin Laboratory. He also had some work going on on the 14th floor; that was a very tall laboratory for those days. And Wally Zinn had a Cockcroft-Walton accelerator, a nice little one, on the fifth floor, which was a neutron source for other experiments. The neutron had been discovered about four years before and there wasn't much neutron physics going on in the world. Columbia had, I think, probably one of the best equipped laboratories. So, I started doing neutron physics. This was another step that was even closer to the eventual path to Los Alamos. I collaborated with Dunning on one experiment and then worked with Hy Goldsmith and a fellow named Julian Schwinger

21

L. Badash, J. O. Hirschfelder and H. P. Broida (eds.), Reminiscences of Los Alamos 1943–1945, 21–40.

on another one. Goldsmith eventually went on to be one of the founders of the *Bulletin of the Atomic Scientists* and, as physicists will know, Schwinger later became a Nobel Laureate for his work in quantum electrodynamics. He was a very young graduate student, I think about 16 at that time, but practically had his Ph.D.

Most of the people at Columbia that were doing research were either members of the staff or graduate students, or members of the staffs of other institutions in New York City who came to Columbia because of its good facilities. But there was one exception to that description of people doing research at Columbia in those days in the person of one perhaps slightly older than most of us. He was certainly very pleasant, but a bit formal in the European manner, and he was never seen doing any laboratory work himself. This was a big contrast to the habits of the rest of us there; even the theorists like Rabi would sit in a corner of the lab and at least whittle while experiments were going on. This 'exception' lived in New York and took an avid interest in the neutron physics of those days. He was obviously very knowledgeable about such work in this country and abroad. His name was Leo

\-bomb history.

a to accept a position at my under-
linois. There I started the construction
uld continue doing neutron physics. I
vas still committed to neutron physics,
d to Los Alamos. Lee Haworth and I
ine, a few theses were done with it
e MIT Radiation Lab and later I went
o. The accelerator sat for awhile and
d to Los Alamos. So it was drafted, as

ate '39 to '40, Szilard phoned me and
bia to help with some very important
, of course, that I was doing neutron
was doing, I'd been so busy building
lk to Wheeler Loomis, who was the
at time. Wheeler had many contacts
e said, "Oh, that stuff is so terribly
poorly organized that I won't let you go; that's all there is to it. You've got to stay here and do your teaching and your research work." Well, that lasted until maybe the fall of '41, early fall when the circumstances changed, both in Wheeler's opinion and the general circumstances. At this time Szilard

→ Construction of Cockcroft Walton machine (neutron physics) $1800 grant.

→ problem to find material to slow down neutrons — graphite, heavy water, metalic beryllium

JOHN H. MANLEY

A NEW LABORATORY IS BORN

I have learned from my historian friends that I am classified as a "living source material." I have never quite figured out where in the hierarchy of honors being "source material" comes, but when one gets to my age, the adjective "living" is encouraging at least.

I hope to describe the background to my association with the Manhattan Project, before discussing Los Alamos. Out of this will come a picture of the small world of physics in the 1930's, where people of my generation knew most of the leaders and had intimate contact with many of them.

I think it's fair to say that the start on the road to Los Alamos was an appointment at Columbia University following my doctoral work at Ann Arbor. At Michigan I'd been interested in atomic physics, and I don't mean atomic bomb physics at all, but physics of the atom and not physics of the nucleus. But, on moving to Columbia, I did start to get interested in nuclear physics and nuclear physics was the basic physics for the bomb. At first when I went to Columbia I worked with I. I. Rabi, making measurements of particular nuclear properties with a molecular and atomic beam method for which Rabi later got the Nobel Prize. My own feeling, I remember at the time, was that I didn't care too much about New York City and that the particular research work Rabi engaged in was too specialized for my taste and not easy to take somewhere else if I were to move. Prompted by this feeling, I started looking around for other things that were going on at Columbia. I had gone there in 1934, and perhaps after a year and a half working with Rabi I shifted into neutron physics. Neutron physics at Columbia was a very active area. John Dunning had some people building a cyclotron in the basement of Pupin Laboratory. He also had some work going on on the 14th floor; that was a very tall laboratory for those days. And Wally Zinn had a Cockcroft-Walton accelerator, a nice little one, on the fifth floor, which was a neutron source for other experiments. The neutron had been discovered about four years before and there wasn't much neutron physics going on in the world. Columbia had, I think, probably one of the best equipped laboratories. So, I started doing neutron physics. This was another step that was even closer to the eventual path to Los Alamos. I collaborated with Dunning on one experiment and then worked with Hy Goldsmith and a fellow named Julian Schwinger

21

L. Badash, J. O. Hirschfelder and H. P. Broida (eds.), Reminiscences of Los Alamos 1943–1945, 21–40.

on another one. Goldsmith eventually went on to be one of the founders of the *Bulletin of the Atomic Scientists* and, as physicists will know, Schwinger later became a Nobel Laureate for his work in quantum electrodynamics. He was a very young graduate student, I think about 16 at that time, but practically had his Ph.D.

Most of the people at Columbia that were doing research were either members of the staff or graduate students, or members of the staffs of other institutions in New York City who came to Columbia because of its good facilities. But there was one exception to that description of people doing research at Columbia in those days in the person of one perhaps slightly older than most of us. He was certainly very pleasant, but a bit formal in the European manner, and he was never seen doing any laboratory work himself. This was a big contrast to the habits of the rest of us there; even the theorists like Rabi would sit in a corner of the lab and at least whittle while experiments were going on. This 'exception' lived in New York and took an avid interest in the neutron physics of those days. He was obviously very knowledgeable about such work in this country and abroad. His name was Leo Szilard, a name you'll find throughout A-bomb history.

In the fall of 1937, I left Columbia to accept a position at my undergraduate alma mater, the University of Illinois. There I started the construction of a Cockcroft-Walton machine so I could continue doing neutron physics. I had a grant of $1800 for this project! I was still committed to neutron physics, which was another factor along the road to Los Alamos. Lee Haworth and I collaborated in completing that machine, a few theses were done with it before the war, then he went off to the MIT Radiation Lab and later I went off to the Metallurgical Lab at Chicago. The accelerator sat for awhile and then in 1943 was dismantled and moved to Los Alamos. So it was drafted, as well as Haworth and myself.

While I was still at Illinois, along in late '39 to '40, Szilard phoned me and asked me to consider coming to Columbia to help with some very important work that was going on there. He knew, of course, that I was doing neutron physics; I didn't know exactly what he was doing, I'd been so busy building the accelerator and so on. But I did talk to Wheeler Loomis, who was the head of the department at Illinois at that time. Wheeler had many contacts throughout the world of physics and he said, "Oh, that stuff is so terribly poorly organized that I won't let you go; that's all there is to it. You've got to stay here and do your teaching and your research work." Well, that lasted until maybe the fall of '41, early fall when the circumstances changed, both in Wheeler's opinion and the general circumstances. At this time Szilard got after me again; I did get leave of absence from Illinois to go, not back to Columbia but to the University of Chicago, to what was called the Metallurgical Laboratory, which was just being set up under Arthur Compton. The work toward a chain-reacting pile, which had been going on at Columbia and at Princeton primarily, but at Chicago also, was being consolidated by Arthur Compton at Chicago in this newly named − deliberately to be misleading − 'Metallurgical Laboratory.' I reported there in January of 1942.

The months spent at the Met Lab were thrilling, hectic and varied. I got back into the swing of the neutron physics that had been going on at Columbia. I was working with old colleagues from Columbia like Szilard, Zinn, George Weil, and Herb Anderson, whom I'd had in an undergraduate course. Herb was now one of Fermi's right-hand men on the chain reaction work. Fermi had come to Columbia from Rome, but after I'd left for Illinois, so I didn't know him from the Columbia days. I'd heard Fermi lecture in Ann Arbor during summer sessions, and had been terribly impressed by the clarity and brilliance of his lectures. I would now have a chance at Chicago to get better acquainted with Fermi and many of the other people who gradually collected at that Metallurgical Laboratory activity. There was Compton himself, Sam Allison, Eugene Wigner, John Wheeler, Bob Christy, Glenn Seaborg, Alvin Weinberg, and many, many others. Quite a number of these later came to Los Alamos. One who did had just received a bachelor's degree in physics and helped me on a couple of jobs at the Met Lab. This was Harold Agnew, who directed the Los Alamos laboratory from 1970 to 1979.

I was put in charge of a group that used the Cockcroft-Walton that Sam Allison had built, and we worked on various experiments that were relevant to the chain reaction. One problem was to find materials to slow down neutrons. The main scheme was to use graphite. Heavy water was also a candidate but there probably wouldn't be enough available in time. A third possibility was metallic beryllium. We obtained some miserable blocks of metallic beryllium and tried to measure its neutron properties. I think the sample was so poor and our measurements not much better that there was not any very sound conclusion from that experiment. Apparently it would be very difficult to obtain beryllium metal even if the neutron properties were all right.

Hardly four months after I came to the Metallurgical Lab I was given another responsibility. The Uranium Committee had given Compton not only the responsibility for making the chain reaction go with the graphite-uranium pile, but he also had under his wing all of the so-called fast neutron work, which was work that led to the bomb itself. This included both experimental and theoretical studies on how to make a fast chain reaction that would really give an explosion. And at the time of the Metallurgical Laboratory's formation in January of '42 he had chosen Gregory Breit to supervise this fast neutron or weapons physics work. After a fairly short period of time, Breit decided to resign and Robert Oppenheimer was Compton's choice to succeed him. Oppenheimer had been guiding the theoretical work of a group in Berkeley and had helped Compton in a report the previous year. Oppenheimer wanted help from someone with more experimental experience than he had and he accepted the position on the condition that an assistant be chosen who would fill in this gap. I may note parenthetically that almost *anybody* would have had more experimental experience than Oppenheimer. He had essentially zero laboratory experience. He was an outstanding theorist and did understand the laboratory techniques, but not as an operator. Anyway, I was Compton's choice. I had gone to Chicago thinking I was going to work on the chain reaction with Szilard and Fermi and company, but that part only lasted four months, although I kept charge of my group and tried to do the other stuff too, which was characteristic of the way things went in those wartime days. I let myself be persuaded to join Oppenheimer with some misgivings. I had only briefly met him. I had given a colloquium in Berkeley a year or two before and I was somewhat frightened of his evident erudition and his lack of interest in mundane affairs. Those things were no assurance to me that I could satisfactorily collaborate with this high-powered theorist and make a go of the weapons end of the thing as it was set up at the time. But for all my doubts at that time, this was really the beginning of an association and a friendship with this truly remarkable man, from whom I certainly learned very much and who had a deep and positive influence on me and my life.

The physics of an explosive chain reaction is really very different from the physics of the chain reacting pile. In the latter, the pile has to operate on slowed down or moderated neutrons, but not so in the chain reaction for a bomb. It must go fast and be propagated by fast neutrons, otherwise the reaction time would be so long that the material would simply expand and you'd get a fizzle with no real explosive result. And so a point to remember is that bomb design depends very critically on the fast neutron properties of all materials that are involved in the bomb itself. In contrast, the crucial thing about the chain reacting pile concerns the slow neutron properties of the materials. Measurements of all these properties of fast neutrons require neutrons of known speed and energy. The accelerators that we speak of, cyclotrons, Cockcroft-Walton and Van de Graaff machines, are just the sources that would supply the fast neutrons for the measurements that were necessary.

When Oppenheimer and I took over the work which Breit had been supervising I had to chase around the country because there were, I think, nine separate contracts with universities that had accelerators which could be used as neutron sources. Measurements of fast neutron properties were made at all of these places — everywhere from Washington, D.C. to Rice to Minnesota, Wisconsin, Purdue and so on. Oppenheimer, on the other hand, had a small group in Berkeley which was concentrating on the theoretical problems and calculating with the data which the experimental programs would feed him. I can't tell you how *difficult* those experiments really were. The amounts of material to work with were infinitesimal. They were separated with Al Nier's mass spectrograph, usually in microgram quantities, just practically invisible quantities. It was hard to assay the samples and hard to make experiments on the targets made from this material. The sources of neutrons couldn't be calibrated very well, so one didn't know how many neutrons one really had and so on. It was a pretty discouraging sort of business, and it was all very new. Oppenheimer would come east periodically to catch up on the latest experimental information, to try to plan where to go next and to decide what measurements looked so miserable they couldn't be used and would have to be repeated, and so on. The problem of liaison among all the groups was a fantastically difficult one. We couldn't, of course, use long distance telephone; our work was classified. Teletype connections that *could* carry classified messages were limited and next to hopeless for trying to unsnarl experimental difficulties among these various groups. I think it was only the very high competence of the people in the various groups and their hard work that made *anything* come out of this period. The point of all this is that it didn't take very long for Compton and Oppenheimer and me to realize that we just couldn't run a railroad in this fashion and get anywhere in finding out the necessary properties of the materials from which to build a bomb. You can imagine how crucial this was, you couldn't even tell how much U-235 or plutonium you'd need in order to make one explosive weapon and all of the production plants had to be designed in connection with *some* estimate of how much material would be required per weapon. Things were really in very much of an uncertain mess in this period of 1942.

We were so upset about the situation that shortly after General Groves was appointed in charge of the whole project we approached him about establishing a new laboratory where one could bring together all these separate groups, have an interchange of ideas on the experimental and theoretical difficulties instead of all this running around the country between groups of theorists and experimentalists. This consolidation was the main reason for Los Alamos. It was just impossible to operate in the dispersed fashion which we had first tried to do. Groves saw the logic right away, not only from the technical point of view, but he also saw it from the security point of view. If he could coop these people up in one place, it would be a lot easier to control their talking. So I think he liked that aspect of the Los Alamos idea, or Site Y as it was then called.

Many people have the idea that Los Alamos was in the desert area of New Mexico. In fact, the school was on a mesa amidst magnificent mountain scenery, about 35 miles from Santa Fe. The school itself consisted of the Big House, a dormitory for the students which was torn down shortly after the war started. There was also Fuller Lodge, still standing with a couple of wings added. It was made into sort of a hotel for visitors. It's a beautiful building. These were situated on a plateau just about timberline, elevation 7200 feet, with the Jemez Mountains behind, rising to roughly 11,000 feet. The area is very nicely wooded. There were fields on the mesa where dry farming was possible. Just behind the Lodge, there was a pond used for ice skating in the winter and canoeing in the summer time. From the porch of the Lodge one could look east across the Rio Grande valley. There's a drop of nearly 2000 feet from the plateau down to the river, beyond which the Sangre de Christo Mountains rise to a height of 13,000 feet about 40 miles away. Clearly, this is very far from a desert; in fact, it is a very attractive place with lots of good places to hike and stroll and scenery to enjoy. The site requirement that Groves added to the others – that his prima-donnas be happy at Los Alamos – was well satisfied.

Los Alamos was the second major site to be selected. Oak Ridge was chosen in September, 1942. Los Alamos in November, and the big site in Washington for the Hanford piles in January 1943. On December 28, 1942, President Roosevelt authorized an expenditure of 400 million dollars for the Manhattan District, the atomic bomb project. The actual expenditure, of course, from the time the Manhattan District was formed until 1946 was 2.2 *billion* dollars. The thinking in 1942 was that 400 million would take care of producing U-235 from separation plants or Pu-239 from chain-reacting piles and pay for weapons from these materials. Did you ever happen to think that the amount of uranium that has to be separated out, that is, the useful isotope for a weapon, is about like the impurity in that famous 99 44/100% pure Ivory Soap? Furthermore, it's a lot easier to get out that soapy chemical impurity than it is to do the isotope separation for roughly the same amount of U-235 that occurs in normal uranium, namely 0.7%. Of course plutonium production had to depend on the chain reaction, but the end of 1942 was a magic period because, on December 2, Fermi and the rest of them in Chicago succeeded in making the chain reaction in the graphite pile go. I happened to be in Chicago since my office was still there, and knew what was going on in the west stands of Stagg Field that day, but I didn't even bother going over that afternoon. Perhaps my failure to attend was a particular quirk of my own, but I like to think of it as somewhat illustrative of the pressure and the tempo of our work. We were all very busy trying to get our jobs done at that particular time in history, and it did not seem sufficiently attractive to go over a couple of blocks to see the chain reaction start. I've been sorry since.

It wasn't very long after the Los Alamos site selection that Groves succeeded in getting the University of California to be the contractor to operate the new laboratory. Today this is the eastern-most campus of the University of California, the Los Alamos campus in New Mexico. It is so because California is still operating the Los Alamos Laboratory.

But what was the concept of this laboratory in 1942? What were we thinking about? Well, we were thinking mostly about neutron physics, the fast neutron physics that had to be done, because that was the bottleneck. That's where we were having all the difficulties with the science involved. We thought if we could get good measurements of nuclear properties it wouldn't be too hard to do the business of getting explosive material together somehow so that it could go bang. But the important questions were how much material could be assembled in one weapon and how much could that amount of material be reduced by surrounding it with some 'nice stuff.' Whatever that nice stuff might be, it would have to do two things: One, hold the explosive together for a while so it wouldn't expand too fast and blow itself apart, stopping the reaction. This means something with high inertia, a heavy material. The second purpose of the stuff would be to reflect back neutrons effectively so that those that *might* escape would come back and feed into the chain reaction and you'd get more energy out. These were problems of neutron physics you see. Another question was what explosive power could you achieve from all of this. The answer involved knowing many nuclear properties and calculating what you might expect the assembly to do. Another very crucial question was how fast would you have to assemble these sub-critical masses together so that you'd really get a bang instead of just a poof, a fizzle. That was a tough question which caused us a lot of pain. You have to remember, too, that it was very likely these questions could not be answered by direct test because there would not be enough explosive, fissile material.

In 1942, the plants were only authorized, certainly not in production. There was no electro-magnetic separation plant, no gaseous diffusion plant. We were guessing about the rates of production, and on top of that didn't know how much was needed to make one weapon, within a factor of 10 let's say. It was felt likely that there would never be a chance to test the explosion of a weapon. What you would have to do would be to make individual measurements and then depend on calculations and theory to say what would happen when you put these critical masses and sub-critical masses together, and go right into combat with that sort of device.

In this situation we planned to obtain two Van de Graaff generators from Wisconsin to do the necessary physics measurements, as well as the cyclotron from Harvard and the Cockcroft-Walton (mine) from Illinois. All of these were shipped to a medical officer in St. Louis and then trans-shipped to Los Alamos. The medical officer, of course, was a blind to try to throw off any interested people from the track of where these things were going. Then we planned to recruit scientists, mainly from the existing fast neutron physics groups, people to man the accelerators and do other necessary experiments at Los Alamos, something on the order of a hundred scientists altogether perhaps. The Ranch School would give us initial housing and more could be built while we got the laboratory started. There was something of a precedent for this since Compton had brought all pile work to Chicago by moving the Princeton and Columbia work there. But this concentration was at a going university, with libraries, machinists, glass-blowers and staff. It was in the middle of a big city with all sorts of supplies. What we were trying to do was build a new laboratory in the wilds of New Mexico with no initial equipment except the library of Horatio Alger books or whatever it was that those boys in the Ranch School read, and the pack equipment that they used going horseback riding, none of which helped us very much in getting neutron-producing accelerators. We had to get all of our needs together from wherever it could be found and ship it to Los Alamos. I assure you it was not a very simple task in comparison with starting up in the middle of Chicago. I've often wondered whether, if Oppenheimer had been an experimental physicist and known that experimental physics is really 90 percent plumbing and you've *got* to have all that equipment and tools and so on, he would ever have agreed to try to start a laboratory in this isolated place. But anyway, it was done, and the laboratory got doing.

I was supposed to talk to people in the fast neutron groups at Princeton and Wisconsin and other places, and try to persuade them to come to Los Alamos. But *I* didn't know anything about Los Alamos. Oppenheimer had told me a few things about it — near Santa Fe and so on — so I dug out some maps of New Mexico and I looked all over those maps trying to find where it might be. He'd said it was near the 'Hamos' Mountains, and I looked for HAMOS and I couldn't find it on the map, *any* map of New Mexico. I hadn't had any Spanish and, of course, I didn't know that those doggone mountains are spelled JEMEZ. That will suggest, I think, the ignorance of this particular recruiter. However, either through patriotism or a sense of loyalty or adventure — I don't know what — most of them were agreeable to take a crack at this unknown, and promised to come to Los Alamos and do the necessary experiments. I did have difficulty with one of the female, unmarried physicists because she thought that her social life wouldn't be very adequate at this isolated place and I couldn't tell her anything differently. She couldn't know that before the end of the war, if she had come (she didn't) she'd have had the company of about 3,000 GI's, most of them single, and only about a hundred WAC's for competition.

In addition to talking to the people about going to Los Alamos, I talked to them about laboratory facilities, what we'd need, what kind of laboratories for the Van de Graaff and the cyclotron and so on. Our ideas were communicated to the architect/engineering firm (Stone and Webster in Boston) which had been selected by General Groves. In December 1942, I went from Chicago to Boston to look over the drawings for the laboratory buildings and to OK their blueprints of the initial plans for offices and other buildings, especially housing for the accelerators. They seemed adequate, except I was worried about one thing, I remember. I hadn't seen the ground on which the buildings would be erected. I tried to protect myself a little bit and also cut construction time by marking on the drawings that the contractor should take advantage of the terrain in locating the buildings. I had particular concern about one very long, narrow building, which was full of laboratories and offices along a single corridor. At the end were two appendages, just barn-like affairs. One would house the two Van de Graaffs and the other the Illinois accelerator. Well, the Van de Graaffs were very heavy instruments and the accelerator from Illinois was a vertical machine which required a basement, so we'd specified that a basement be excavated for that machine and that there must be a good foundation under the Van de Graaff accelerators. Cost and construction time could obviously be saved if they selected the terrain properly. I shall return to this shortly.

There was another item which worried me very much and that was the organization of the laboratory. There were several reasons for concern. People would tire in a very isolated place. They would be working under extreme time pressure, and if there was not a good organization from the point of view of the technical work, all of the services, the responsibilities, the whole enterprise could just really go flop under those pressures. I had another motivation too. I was trying to tell people about Los Alamos, to get them to go there. It's awfully hard to convince somebody to move to a place if you can't tell him something about how it's going to be organized. I bugged Oppie for I don't know how many months about an organization chart — who was going to be responsible for this and who was going to be responsible for that. But each time he would seem to be about as unresponsive as an experimental physicist would think a theorist would be, and I'm sure he was, maybe more so. But one day in January — I remember very vividly because I had taken a DC-3 flight out of Chicago, first to Denver and then out of Denver we ran into a blizzard and went up to I think 17,000 feet to be sure we cleared all of the mountains, and I didn't get the oxygen tube in time (those old things that you had to stick into a little plug in the wall, you remember on a DC-3?) We couldn't land in Salt Lake, so we landed in Reno, the storm was so bad. Anyway, I finally got to Oakland and then over to Berkeley and climbed to the top floor of LeConte Hall where Robert had his office, and pushed open the door. Ed Condon happened to be in there with him at the moment, but Oppie practically threw a piece of paper at me as I came in the door and said, "Here's your damned organization chart." Well, that was the way the lab was organized to start with and the initial organization sort of stuck that way with one exception. He had kept for his own responsibility being head of the theoretical division. It didn't take him very long, especially with some pretty good pressure from I.I. Rabi and also from Bob Bacher, to realize that he couldn't manage the laboratory and also look after all of the theoretical work. He gave in and appointed Hans Bethe to take care of the theoretical division work at Los Alamos.

In April of 1943, people started coming into Santa Fe, reporting to Dotty McKibben at 109 East Palace, asking for further instructions: where do they go from there? She'd tell them how to cover the last leg, which was that 35 mile trip. I had to stay a few days when I arrived in Santa Fe just about April Fool's Day (maybe quite appropriately) in order to get the accelerator from Illinois unpacked. It was in the freight yard, in a boxcar and I had to make arrangements to get it on a truck and up to Los Alamos. I remember it was April fourth when I drove off with a man and his truck, and the accelerator in the back end of the truck, to my unseen destination of Los Alamos. I was quite concerned most of that 35 mile drive for the cargo in the back of the truck. The tube of the accelerator we were carrying was made of several procelain crocks waxed to metal plates between each pair. In our haste to get everything crated and shipped from Illinois, we just clamped the tube together with a 6 or 7 foot rod down the middle. Now, the crate holding this tube, vertical and without stays in the back of the pickup, was waving precariously on every turn. It was a steep old country gravel road full of turns and sharp bends as it climbed the last 12 miles from the Rio Grande crossing at 5500 feet above sea level to the 7200 foot altitude of Los Alamos. We managed to make it and pulled up in the laboratory area. Of course the first thing I wanted to see was the building that I'd specified be oriented properly back in Boston. There are enough jokes about the way of the Army so you can guess what I saw. The basement for the Illinois Cockcroft-Walton had been dug out of solid rock and that rock debris taken over to the other end of the building and used for fill under the Van de Graaffs, where there was supposed to be a good foundation. This was my introduction to the Army Engineers.

I think we really must have made some sort of a record and I'm still moderately proud of the little part I had in it: all four of these accelerators — the cyclotron, two Van de Graaffs and a Cockcroft-Walton — were operating and giving data on experiments within two or three months after they were trucked up to Los Alamos. This would have been a feat on any university campus, but at Los Alamos the buildings weren't even finished, the workmen were still in there, special wiring and plumbing had to be installed for all of these machines, and shops and stocks and everything else were just getting organized. With luck, a missing piece of electronic gear that was supposed to have been shipped, say, from Princeton, and didn't arrive might be picked up in Albuquerque or in Santa Fe if it were simple enough. But usually it meant a very involved procedure. What you had to do if you wanted to buy anything was to get a requisition; if it were a rationed item, you had to get a priority. There were three purchasing offices, one in Los Angeles, one in Chicago, one in New York. They went out and tried to buy it and then disguise the bill of lading and the shipping so that nobody could ever figure out who had bought it or where it went. It was good luck, after all that procedure, when the item really arrived back at Los Alamos. I can add to the picture of our early complications by reminding you that the only telephone line to Los Alamos from the outside world was a Forest Service line. If you manipulated the crank on the side of the box with enough vigor sometimes you got a response from the outside world.

If there were any ground-breaking ceremonies at Los Alamos like champagne or cutting ribbons, I was unaware of them. Most of us who were there felt that the conference in April, 1943, was really the ground-breaking ceremony.

To that conference came practically every distinguished scientist who had any inside part of the uranium project. It was a complete going-over of all the possibilities for the weapons and a serious effort to lay out a sensible program – what measurements were needed and how they would be accomplished. It would be too technical to try to summarize that conference and I won't do so, but I might mention one assignment. It was the one to my own group, which rather naturally drew the responsibility of operating the accelerator which I'd built at Illinois: we were charged with trying to find the best tamper material. Now that word, tamper, comes from ordinary prose. One tamps dynamite in a hole to confine the explosive force. For the bomb, the tamper would surround the explosive fissile material, kind of hold it together, and reflect back the neutrons.

It is hard to re-create the atmosphere of those days. Let me try with an illustration. Just before Los Alamos really got going, the last measurements on how much uranium 235 might be needed for a weapon had increased over the previous low estimate by almost a factor of 2; it was about 5 kilograms in absolute amounts. These 5 kilograms meant nearly two months extra production for each weapon from the electromagnetic separator which had been authorized at a hundred grams per day. Since we had no idea where the Germans were in this whole business – whether they had isotope separation plants going, whether they had a chain reaction going and were making plutonium, or were almost ready to drop bombs – these two months could mean we'd lose. However, there was a chance we could recover some of this apparent loss. Maybe, if we were really clever and got an extremely good material that could reflect back neutrons and behave properly, we could get back most of that factor of 2 that we had just lost. We were playing that kind of a game almost continually. You get experimental results which say things are worse than expected. Then you try to be smart and get some better material or a new device or do things another way.

I hope these comments convey a correct impression of the beginnings of this new laboratory and its people. Perhaps it could be called a new civilization colonizing this Pajarito Plateau of northern New Mexico, some 800 years after the first-known permanent inhabitants of this particular region, the Keres people, came to this plateau about 1150. This re-colonization was for a very different purpose than the earlier one. Our task was to pursue a development to resolve a conflict of half the world. To make a new scientific laboratory and a new community was a step in that task.

In closing, let me discuss three topics that I sense will answer some obvious questions. The first relates to why the initial size concept of Los Alamos was so far off. Originally we thought in terms of 100 scientists and obviously that wasn't a good estimate. Well, of course, you could say we were just not far-sighted enough, but I'd like to mention a few things that were really extenuating circumstances. One was the matter of plutonium fabrication. We did plan to have a chemistry and metallurgical group at Los Alamos but on a laboratory scale. Groves had a habit of using advisory committees throughout the whole operation and he had one come to Los Alamos to look things over. The committee advised him that Los Alamos should take on the final purification and fabrication of plutonium into metal for *all* of the Hanford output. That added a considerable extra effort to the Los Alamos activity in terms of people and plant. It was a logical decision though, because the material was never very abundant, and if you did one experiment and a following test had to have a different shape, the material would have to be reworked. It made lots of sense to do it right at Los Alamos. One really couldn't quarrel with that decision. Then there turned out to be some problems with assembly, as is well known. We thought we could just go to the military and buy a gun that would blow a couple of pieces together fast enough to make an explosion. But fast enough turned out to be really very fast. On top of that, the whole business had to be carried by a B-29 and dropped as a ballistic missile and the Navy or Army just don't make guns for those purposes. All of this put very stringent size and shape and weight requirements on a gun. The upshot was that for the most part the gun was designed and tested at Los Alamos. Again, that took more people.

The really big jolt occurred in July 1944. This terrible shock, and an inescapable one, was that the gun assembly method could not be used for plutonium. Of the two fissile explosives, U-235 and plutonium, we finally had to conclude that a gun just would not assemble plutonium fast enough. Another isotope besides the one that we wanted was also produced all the time in the piles. One could have separated out those bad plutonium isotopes from the good ones, but that would have meant duplicating everything that had been done for uranium isotope separation — all those big plants — and there was just no time to do that. The choice was to junk the whole discovery of the chain reaction that produced plutonium, and all of the investment in time and effort of the Hanford plant, *unless* somebody could come up with a way of assembling the plutonium material into a weapon that would explode. It was again a Los Alamos challenge to do just that. It was the only possible way out at that stage of the game. Now, fortunately, a fellow named Seth Neddermeyer, a physicist, had been playing around, experimenting with an idea. He had seriously proposed a new way of assembling a weapon and though nobody else really took it very seriously he kept on working with it, from the time Los Alamos started in early '43. When this difficulty came up in '44, his work was really a godsend. He had carried it on almost by himself, but now everything changed and he got lots of support. The new assembly method, which is now known as the implosion method, was developed and made into a successful technique by which plutonium could be used. This too had an extra effect on the manpower situation.

Finally, we had never really intended to test the uranium bomb. It was a gun assembly, the components could be thoroughly tested, and we felt quite sure it would work. There was not enough fissile material to spend on a test anyway. This bomb had to go right into combat. However, the implosion scheme was so new and so complicated to predict what might happen that it was decided a field test would be absolutely essential before combat use. That was the Trinity Test, near Alamogordo in New Mexico. It was a tremendous field operation which took lots of people, lots of ingenuity and so on, but it was accomplished. That again, added to the number of people involved in the Los Alamos enterprise toward the end of the war. The value of this achievement in which we proved the ability to use the plutonium as a nuclear explosive is nicely illustrated by some comments from Groves' book. He points out that there was never enough, could not be enough, U-235 to be able to afford a test of it; its production was so slow compared with plutonium that that itself was reason for not having a test. Plutonium's rate of production was much better than U-235, and its relative abundance permitted testing the plutonium implosion system on July 16, 1945. The plutonium bomb was dropped on Nagasaki on August 9 and another such weapon was ready on August 24. So the plutonium production was coming along, producing bombs at a respectable wartime rate, if you can talk about such things being respectable. By May of 1945, Los Alamos had expanded to about 1100 civilian employees and over 1000 Army technical people. In spite of many difficulties, the scheduling of the whole project was never held up by a weapons problem; it was always the production of materials for weapons that determined the time scale.

One could just as well invert my question and ask why was it that Los Alamos was able to accomplish this remarkable task with no *more* people than it had? An example from my own group may help to understand why. We got the work on the tamper material pretty well in hand by the spring of '45 and that meshed very nicely with the necessity to test the implosion device that summer. Although most members of my group were nuclear physicists, they also had solid academic backgrounds of teaching physics.

We did not hesitate, therefore, to accept the assignment to conduct the blast measurements at Trinity. This was a direct way to determine the magnitude of the explosive force. Using basic physics principles, reading the literature, nabbing a few experts, and talking to the theorists, we became instant blast experts. We set up a shock tube in the laboratory next to the Cockcroft-Walton machine to calibrate instruments and do experiments. I even had to go out and run a surveyor's transit myself because there were no cleared people who could help lay out the field stations. I knew how to do trigonometry, I could read a tape, and I could level up a transit, and that's about all there is to the business of surveying. In this way it got done. This example is illustrative of a process that went on continually at Los Alamos, namely of shifting people and responsibilities to the job as it occurred. This could be done because of a breadth of competence of most of the staff. We would have been sunk if we'd had to hire new people, get clearance for them, bring them in, find the housing for them, and so on. It never would have worked. Staff flexibility was a very important factor at Los Alamos.

My final point concerns secrecy. I've already mentioned that there were difficulties of procurement and recruiting *because* of secrecy and security. It was a continual nuisance, no doubt about it, but I think because it was such an obvious requirement that it was really quite acceptable. The remarkable way in which the purpose of the whole operation was kept from public view seems to me to attest to the fact that people really *were* cooperative about the necessity for secrecy and approved generally of the way the security operation was run. In one area, that of technical discussions between scientists, the normal security procedure of limiting topics to a 'need to know' was not followed. Oppenheimer adamantly refused to permit any compartmentalization. Each scientist could discuss his work with his peers. Groves acceded to this policy, for he found Oppenheimer strongly supported by Conant, Rabi, Bacher and all the rest of us. A feature of Los Alamos routine included regular formal colloquia and many informal meetings. These discussions brought forth ideas and were excellent morale boosters. I'm sure the work went faster and more effectively as a result. After all, a reason to create Los Alamos was to improve the communication between all those engaged in the weapon problem.

Another observation on secrecy concerns a personal experience just before the war ended. The advance base for operations against Japan was the island of Tinian. However, there was no direct communication link from there to Los Alamos; messages had to be relayed through Washington. Oppenheimer was a little worried about this extra link, for he wanted to be in close touch with the Los Alamos people on Tinian who had the responsibility of readying the bomb and loading it in a B-29 for the trip to Japan. As a result I drew an assignment to Washington as a one-man liaison office for Los Alamos. My job was to try to keep everything flowing smoothly through this communication link in Grove's office. We had no trouble until the very last minute — at attack time essentially. In Washington, the date was August 5, 1945, a Sunday. We knew on that Sunday when the strike plane was scheduled to take off (noon Washington time: about 2:00 a.m. in the far Pacific) and we were supposed to be in continuous communication with Tinian from then on. But Tinian was silent all afternoon. We heard nothing at noon as we had expexted, in fact the takeoff report didn't come through until almost 7 o'clock that evening. I was in Groves' office and we were all pacing up and down. Groves went off and played tennis; that was like him. The time that the report of takeoff came was just about the time we should have been receiving the report of the effect of the strike over Hiroshima. There was more fidgeting, but no more reports; we knew only that the plane had left on time. Where was it? We didn't know. Well, about 11:30 at night we finally got the report that we'd been expecting almost 5 hours earlier. It was the message from the strike plane, sent by Captain Parsons of Los Alamos who was on the aircraft. He gave a visual description of how the explosion over Hiroshima appeared to him. From all he could tell, it was successful. Here was the culmination of the Einstein letter to Roosevelt, of the Met Lab, Oak Ridge, Hanford, and Los Alamos and we were having a 5 hour delay in our communications between Washington and the operational area! It was exasperating!

But this was just the beginning of my own exasperation that night and the next day. Here I sat in Groves' office with meagre but vital information about the successful strike — Parsons had said, "the visible effects were greater than the New Mexico test." I was where I was to provide the liaison with Los Alamos. This was the biggest news that we ever had in the whole project, and the reason why we'd all worked so very hard. Then, as I started to relay the news to Los Alamos, Groves told me that I couldn't do so. He absolutely refused to let me get through to Oppenheimer on the teletype, the telephone, or anything. Why? Because there were military orders which he said had been issued. Oppenheimer and I didn't know about those military orders. Later I learned that they contained a sentence which read, "Dissemination of any and all information concerning the use of the weapon against Japan is reserved to the Secretary of War and the President of the United States." And this was signed by the Secretary of War! Well, Groves took the position that even Oppenheimer couldn't warrant violation of these orders and that was that.

I wasn't kept locked up in Groves' office that night, and I decided to take a break and go for a long walk from his office, which was in what is now the State Department Building on Virginia Avenue. I walked all the way down Pennsylvania to the White House. Lafayette Park was still pretty active with soldiers and sailors and their girlfriends — this was probably one o'clock in the morning. I couldn't help but thinking: here you are, fully expecting that you are going to go back on duty as soon as this leave is over, and probably have to go and invade Japan. If *you* only knew what *I* know right now, and I can't tell you about it! It was a strange feeling to have this clamp on me. I even toyed with the idea of stepping into a phone booth and calling Oppenheimer long distance but somehow I just couldn't do it. We had prearranged a code but I was afraid it would be broken, all hell could break loose, and both Oppenheimer and I would be in real trouble. I decided to be a good boy, went back to the office and spent my time summarizing later reports as they came in for the teletypes that would go out to Los Alamos just as soon as Groves would release them from his office. We got word early that morning that President Truman, who was on the cruiser Augusta, coming back from Potsdam, would make the formal announcement about the dropping of the bomb on Japan at 11 o'clock, Washington time. Just a few minutes before 11, I got Oppenheimer on the long distance phone and as he was hearing the President, he was also hearing from *me*. The first words that he said to me were not in any pre-arranged code and they were very clear: "Why the hell did you think I sent you to Washington in the first place?" Well, there's an epilogue to this little tale of communication. It concerns the teletype that I finally sent from Washington and the way it was received at Los Alamos (see pp. 39–40). It begins, "Flashed from the plane by Parsons one five minutes after release and relayed here was this information." Note the horrible mess into which the text degenerated. That's the way in which the director of the laboratory that designed and built the first atomic bomb was officially informed of its combat use, by way of a fancy, secure teletype circuit from Washington. I've often wondered if that teletype machine was trying to tell us something about science and technology and human affairs.

DISCUSSION

Question: Why was it apparently so easy for Klaus Fuchs to penetrate Los Alamos' secrets and send them to the Soviet Union?
Manley: In a certain sense he didn't *have* to penetrate. He was an official member of the British mission. For instance, I worked very closely with him

because he did a lot of the calculations estimating the blast, calculating curves of the blast as a function of distance, which we needed in order to calibrate our meters and try to get them on scale for the Alamogordo test. So he was a very respectable member of the theoretical group.

Question: Weren't satisfactory security checks run on such people?
Manley: That depends on who you ask. If you were to ask Groves that question, for instance, he would complain that the whole British mission and even the Canadians, I think, were not carefully enough investigated. And he was suspicious of them. But, like many other people, Fuchs had the same sort of credentials that seemed appropriate. The British security, I learned afterwards, didn't know of his early Communist affiliations in Germany. But remember that he returned to England, and worked at the big atomic research station at Harwell, where he was head of the theoretical division, for 5 or 6 years after the war. Very interesting. I had a very pleasant dinner with him in Abbington at his 'digs' one time. Very pleasant person.

```
NR 137
FROM WASH LIAISON OFC WASH DC AUG 490C 0402
TO    COMMANDING OFFICER CLEAR CREEK
FIVE PARTS - TART ONE
SW
KC


FLASHED FROM THE PLANE BY PARSONS ONE FIVE MINUTES AFTER RELEASE
AND RELAYED HERE WAS THIS INFORMATION QUOTE PAREN REF EIDM WL
TO OPPENHEIMER FROM GENERAL GROVES THIS RESUME OF MSSSAGES PREPARED
BY DOCTOR MANLEY PAREN CLEAJ CUT RESULTS COMMA IN ALL RESPECTS SUCCES
FUL PD EXCEEDED TR TEST IN VISIBLE EFFECTS PD NORMAL CONDITINXXXXX
CONDITIONS OBTAINED IN AIRCRAFT AFTER DELIVERY WAS ACCOMPLISHED PD
VISUAL ATTACK ON HIROSHIMA AT ZERO FIVE TWO THREE ONE FIVE Z WITH
ONLY ONE TENTH CLOUD COVER PD FLACK AND FICHTERS ABSENT UNQUOTE AFTER
RTXXXXX RETURN TO BASE AND GENERAL INTERROGATION FARRELL SENT THE
FOLLOWICXXXX FOLLOWING INFORMATION QUOTE ALARGE OPEMING IN CLOUD
COVER DIRECTLY OVER TARGET MADE BOMBING FAVORABLE PD EXCELLENT RECORD
REPORTED FROM FASTAX PD FILMS NOT YET PROCESSED BUT OTHEM OBSERVING
MEMBESOALSO ANTICIPATE GOOD TREXXXX RECORDS NXX PD NO APPRE
JQXD   JCFA
R NIL
K   HOW MANY LINES DID U GET
R 12 LINWA
PLANES ALSO ANTICIPATE GOOD RCXXX RECORDS PD NO APPRECIABLE NOTICE OF
SOUND PD BRIGHT DAYLIGHT CAUSED FLASH TO BE LESS BLINDING THAN TRPXXX
TR PD A BALL OF FIRE CHANGED IN A FEW RECORDS TO PURPLE CLOUDS AND
BOILING AND UPWARD SWIRLING FLAMES PD TURN JUST COMPLETED WHEN FLASH
WAS AXXX OBSERVED PD INTENSLY BRIGHT LIGHT CONCEALED BY ALL AND RATE
OF RISE OF WHITE CLOUD FASTER THAN AT TR PD IT WAS ONE THIRD GREATER
IN DIAMETER REACHING THIRTY THOUSAND FEET IN THREE MINUTES PD MAXIMUM
```

ALTITUDE AT LEAST FORTY THOUSAND FEET WITH FLATTENED TOP AT THIS
LEVEL PD COMBAT AIRPLANE THREE HUNDRED SIXTY THREE MILES AWAY AT
TWENTY SIX THOUSAND FEET OBSERVEDIT PD D
NIL AGN
.. OK OPR WELL JUST HAVE TO KEEP TRING AS THESE MESSAGES AR IMP
MIN PLS
OPR U STARTED THIS MSG AS PART TWO ISNT IT PART OF PART ONE
    M   MIN OPR I TOLD U I WD START PART TWO WHERE PART ONE NILED
   IS THAT CLEAR

BUT OPR I DIDNT GET PART ONE COMPLETE

AND THE I TOLD TO U TO SA START WITH 12 LINE
AND THE 12 LINE U L   O WELL I THOT U MEANT U GOT 12 OK
M   THIS IS A AWFUL MESS ISNT IT IT SH SURE IS DOU THINMI   UNGEFG

MIN PLS
TRY ANOTHER MACHINE MAYBE IT WILL DO VETTER
OPR IT ISNT UG MACH AND I KNOW IT   IT S MINE AND THERE ISNT
A THING CAN BE DONE AS THE REPAIR MAN SAYS THERE ISNT ANYTHING WRONG
WITH IT HES BEEN HERE ALL DAY AND THIS IS AS GOOD AS IT WAILL RUN
I HAVE LOADS TO GO UXX TO U TONIGHT BUT WELL HAVE TO DO IT THIS WAY
A FEW LINES AT A TIME   MIN I WANT TO TALK TO THE LT A MIN
OK
   OPR ILL CALL U BACK IN A BT 10 MINUTES
..OK

ELSIE McMILLAN

OUTSIDE THE INNER FENCE

During our first month in New Mexico, in the Spring of 1943, we stayed in a plush ranch in the Nambe Valley because the house we were assigned at Los Alamos was being used by young bachelor workmen and we could not move in yet. On our trips to the site we would drive past the Black Mesa, which seemed to me like a wonderful sentinel, guarding the Indian pueblos so near. The chamiso and the tumbleweeds going along made this wonderful country even more picturesque. We would cross the Rio Grande River on the Otowi Bridge, and ascend into spectacular mountains, with sheer drops, interesting formations, little buttes, and mesas. Only the terrifyingly narrow road detracted from the view.

At the guard gate, we all showed our passes, though of course we didn't need one for our two month old baby. On one occasion, so the story goes, Oppy came up to the gate and went whizzing through. He had a lot on his mind and just went full speed ahead; when the guard shouted at him he didn't pay any attention. Finally when his tires were shot at Robert backed up and pulled down his window and handed a crisp dollar bill to the guard saying, sorry sir.

On the site we saw the stable, where we could take horses out on Saturday or Sunday. There also was the theater, which we could attend for a quarter. On Sunday they would bring in a great big trunk which opened up into an organ which was used for church services. Behind a fence was the technical area, which I couldn't enter since I didn't have a white, pink, or green badge. Among the old school buildings were the Big House, which formerly held the school library and some classrooms, and the Lodge, with a great log front. The water tower was a useful landmark; if I ever got lost I would look for it.

We lived on Bathtub Row, so named because only these eight or so houses had such luxurious plumbing. I hope others weren't jealous; we came so early, that's the only reason we had a former master's home. I didn't like that darned open ditch nearby. It wasn't illuminated at night, and later, two ladies fell in our ditch; they didn't like it either. Our kitchen was very small, and at first I might have bitched a little about it, but then I was happy because my husband said if it was a big kitchen I'd have had a 'Black Beauty.' That's what

41

L. Badash, J. O. Hirschfelder and H. P. Broida (eds.), Reminiscences of Los Alamos 1943–1945, 41–47.

they called the wood and coal stoves, which were in most other homes. But at first we didn't have anything to cook on. There was no room for a coal stove, but they brought in a three-burner kerosene stove. For a while I cooked over the fireplace in our beautiful living room. I had to be most careful with the baby's formula and, of course, things don't cook fast at 7,300 ft. From our porch we could look over the lawn and trees to the Sangre de Cristo Mountains. In summer the pale green of the trees showed a shape called the Thunderbird. In the winter, when it snowed, and the snow somehow appeared in relief, we saw the Thunderbird in white.

One night I said to my husband, "why didn't you tell me you're making an atomic bomb?" He said, "my God, where did you hear that? Do you know you could get me fired?" I replied, "somebody told me but I'm not going to tell you who it was." I am very grateful that I was one of the few wives who knew it was an atomic bomb because I could better understand when my husband left me for places unknown, when he worked all hours of the day and night, when he looked so drawn, tired, worried. On occasion I would partially sleep, and get up and cook another meal at three in the morning.

The commissary on the site, which was quite good, was the only place to buy food. But we needed our food stamps, just as we also required gas stamps, as people throughout the country did. The red stamps, which were for meat, unfortunately in our case also had to be used for our baby's canned milk. As a consequence we didn't have much fresh meat. The commissary had beautiful meat, but their vegetables were terrible.

Returning from the commissary one day I found a soldier standing guard at my door. He saluted me and said, "Good afternoon, Mrs. Oppenheimer, the baby you left in the bedroom is quite all right." I replied, "Thank you very much, but I am not Mrs. Oppenheimer and I didn't leave a baby in my house." He said, "My God, I'm guarding the wrong house!" (We were next door neighbours.) Very shortly after that a fence went up around the Oppenheimer's home. The guards then had to patrol around the fence, and Kitty Oppenheimer and I felt very sorry for them because in winter at Los Alamos we had very dry cold weather; cold enough for icicles to reach from the roof of our one story house to the ground. Kitty and I would sneak out and leave thermos bottles and sandwiches for the guards.

Los Alamos had a small hospital with only two bedrooms, a waiting room, a pharmacy, and an operating room which at first could not be used because we did not have the proper anesthetic for the altitude. We had only two doctors, James Nolan, who had been a cancer specialist, and was the physician for the people who were there (as you remember there were initially maybe a hundred of us, and no one realized how big it would become), and Louis Hempelmann, who was the radiologist for the technical area. We had three nurses, Pete and Sarah and Peggy. Food had to be brought over from the lodge, and it got cold. I was grateful that it was a very unusual hospital because once I was a patient when Ed was away, and because we had no help at home the baby came to the hospital with me, and also Borrego Mac, our tri-colored cocker spaniel. Dr. Jim wondered why my chart read 'thinner and thinner', while Borrego Mac was getting fatter and fatter. On March 25, 1945, Ed and I had the great joy of having our first son born in the hospital. By then the hospital was quite large; I think there were seven rooms. And Henry Barnett had joined us; he was a pediatrician. When our baby was born, or any baby at Los Alamos, the parents immediately got the birth certificate because that meant more red food points, and as I have mentioned, we needed those for the milk. The birth certificates of all babies born there said: place of birth, Box 1663, Sandoval County Rural. My God, was that box ever full of babies!

In age of the residents Los Alamos resembled a college campus. Ed and I were the oldsters; I had my 30th birthday up there. I don't think I shall ever again live in a community where so many brains were, nor shall I ever live in a community so confined that visitors expected us to fight with each other. We didn't have telephones, we didn't have the bright lights, but I don't think I shall ever live in a community that had such deep roots of cooperation and friendship.

Even before the end of our first year at Los Alamos, the emotional strain was apparent, the feeling that you've *got* to make that bomb, you've *got* to get it done; others are working on it; Germans are working on it; hurry! hurry! hurry! This is going to end the war; this is going to save our boys' lives, this is going to save Japanese boys' lives; get that damn bomb done! We were tired, we were deathly tired. We had parties, yes, once in a while, and I've never drunk so much as there at the few parties, because you had to let off steam, you had to let off this feeling eating your soul, oh God are we doing right? They soon realized that we should take a week's trip every now and then to try and relax. On one such occasion we went to Carlsbad Caverns, and on the way drove over the pass from Artesia to Alamogordo. At the very top of the snowy pass we stopped to view the little town of Cloudcroft, and as we slowed the car to gaze at this beautiful sight, out of the woods came the most beautiful, majestic wolf. He stood and looked at us, just stood and looked. I shall never forget that. We went on down to spend the night at Alamogordo and the next day we went to the White Sands National Monument, where we cavorted around the white sands, we wrote on the sand, we rolled down the hills, we had a wonderful time.

As time marched on, Los Alamos got more and more people, more and more buildings. Eventually they covered up the ditches at night, or at least lit them up so you didn't fall in and wreck your clothes or break your ankle. I could go to the commissary and not know some of the people. And the authorities realized that we really needed some help, so the housing office, which had been run so beautifully by Vera Williams, John Williams' wife — many wives worked — was changed to be the maid service. I won't go into the arguments of priority for maid service, but Vera had a very simple system. A sick person got first priority, a full time working wife next, and so on. Ed and I were very excited, as we were to have one day a week of help. There was a knock on my door and Vera Williams came with Pascualita. Pascualita was from the pueblo San Ildefonso. We looked at each other. My great, big, beautiful Pascualita. She wore loose dresses with a wide colored band across her chest; her braids had something I think is very popular with young people today — she used pieces of colored wool entwined in them, and so did the pueblo men in their braids. We were very fortunate because although she worked for four or five households, we were 'her' family. Many a Friday (our day for help) her husband Mr. Peña would come at lunchtime and sit with our Ann on his lap and let her play with his braids.

I will never forget the day Pascualita did not show up. There was a knock on my front door and there was Juanita, her aunt, and she said, "Pascualita no come, she had baby yesterday." Did I feel guilty! I knew she wore loose dresses but I had no idea she was pregnant. Two days later there was a knock on the door and there was the same Juanita. She said, "We have christening. Baby named for you." I said, "Oh! what is the baby called?" And she gave a long Tewa name. "Please, what does it mean in English?" She replied, "Big White Mountain – – – because Pascualita say she see big white mountain from your window."

We had known Indians and been to their dances, but because of Pascualita we had the unique experience of going down to the pueblo and having lunch with the Peñas many times. There were two sides of the pueblo, the poor side and the rich side. She was from the simpler side, which cared a little more about keeping the old Indian customs. They all cooked their bread in the beehive ovens, and made their pottery, the well known black San Ildefonso. On one lunch visit we entered the big room. My namesake was in a suspended bed, like a swing, and the other nine children and the wonderful old pock-faced grandfather sat at the table set with a clean oil cloth. The food was wonderful chili and salads. There was love in that room. There was intelligence in that room. We again noticed that the Indian children never seemed to need discipline. They have, I think, a great closeness to their parents, who do not raise their voices to their children. Two special things happened that day. They presented our two year old Ann with something Pascualita said was the only one in the world, a tea set of black San Ildefonso pottery, which she still has. Little delicate cups and saucers, tea pot, creamer, sugar bowl, and a plate. The other thing that happened that day brought tears to my eyes: they asked Ann to join the Indian children in the corn dance. In all her young life, we never bought a pair of shoes for her because the Peña family, in their generosity, their love, made white shoes from deerskin for Sunday, and brown leather shoes for everyday. What a privilege!

We later had a Spanish-American girl sent to help us. Her name was Frances Gomez, and she and others with very great feeling said, "We are Spanish-American not Mexican; we are proud of being American." We went to her home for her 20th birthday, and the most amazing thing happened. At this birthday party, when it came time to eat, the men went in first, and we women, even the birthday girl, couldn't touch the food until after the men had finished eating.

By the summer of 1945 we all felt that a crisis was pending. Nerves were on edge. The seasonal thunderstorms had started. Each afternoon the heavens would open up. Inwardly I quaked as darkness descended. Lightning came in great jags down the sky; thunder roared in all its fury. I rocked the newborn baby in his carriage and joked with Ann, hoping my fear of storms and the mounting tension would not be transferred to them. Our suspicions were justified. Things were moving fast now. There soon would be a test near Alamogordo, the very place we had visited with carefree abandon earlier. I asked Ed in all innocence what would happen. It seemed an easy question, with a simple answer; knowing that it was an atomic bomb they were testing should have made me more aware of what would be involved. It was difficult for Ed to tell me. He finally answered, "There will be about 50 of us present, key workers." "What do you think will happen?" "We ourselves are not absolutely certain what will happen. In spite of calculations, we are going into the unknown. We know that there are three possibilities. One, that we all be blown to bits if it is more powerful than we expect. Two, it may be a complete dud. Three, it may, as we hope, be a success, we pray without loss of any lives." Unknown to most of us, many dedicated men had been setting up the experiment near Alamogordo for many months. This, then, was part of the reason our men had disappeared periodically on trips, destination, to wives, unknown. Ed continued, "Next week we will quietly and separately leave the mesa, the cars to reconvene at the test site. In all probability, the zero hour will be very early the next morning. If all goes well, I will be home sometime in the early evening of that day. Be sure to look out of the baby's window toward Alamogordo. You may see the flash. I'm not certain, but I suspect you will in spite of the fact you will be about two hundred miles away from it."

That last week in many ways dragged; in many ways it flew on wings. It was hard to behave normally. It was hard not to think. It was hard not to let off steam. We also found it hard not to overindulge in all natural activities of life. Late one afternoon, Niels Bohr, my brother-in-law Ernest Lawrence, General Groves and Bill Laurence, the only member of the press allowed to attend, arrived at Los Alamos. Ernest came home with Ed for dinner that night. Thank God my sister did not know where he was or why he had come. It was no surprise to me when he left early. Ed and I also retired with our alarms set for 2:30 a.m. Ed would leave at 3:15. We did not want to allow much time. I would cook him a hearty breakfast and hope he could eat it. We did not want to say goodbye. Soon he had gone. I was so cold. I was so scared. It seemed long to wait all that day until the early morning of the next before there would be any hope of news. I had to try to get some more sleep. I had to feed the children when they awoke. I had to walk that mesa and appear my usual vivacious self all the coming daylight hours, for many did not know and must not suspect. I prayed in that early morning light. I repeated the Lord's Prayer, especially the phrase "Thy will be done". Somehow the day passed. The children were tucked in for the night.

There was a light tap on my door. There stood Lois Bradbury, my friend and neighbour. She knew. Her husband was out there too. She said her children were asleep and would be all right since she was so close and could check on them every so often. "Please, can't we stay together this long night," she said? We talked of many things, of our men, whom we loved so much. Of the children, their futures. Of the war with all its horrors. We wondered how the people from the technical area, who were keeping watch on the cold Sandia Mountains, were faring. They, too, had given much toward this test. But it was understandable why only a few could actually be on the spot. Those who couldn't had been given official permission to go up the mountain as inconspicuously as possible, to wait for that light in the sky. Lois and I must have consumed gallons of coffee that night. We looked out the back window and silently watched the sky. Nothing but blackness confronted us. We were in Dave's room. He awakened and wanted a bottle. Lois watched out of the window as I heated the bottle as quickly as possible in the kitchen. It was 5:15 a.m. and we began to wonder. Had weather conditions been wrong?

Had it been a dud? I sat at the window feeding Ed's and my baby. Lois stood staring out. There was such quiet in that room.

Suddenly there was a flash and the whole sky lit up. The time was 5:30 a.m. The baby didn't notice. We were too fearful and awed to speak. We looked at each other. It was a success. We had hours to wait to be absolutely sure. At least it was over with. Lois went home to grab a few hours of rest before her family might awaken. I too crawled into bed, but found I could not sleep. The day dragged on. I tried walking the mesa with the children but by lunch-time, home was where I wanted to be. The door opened about 6 o'clock in the evening. We were in each other's arms. Then, and only then, did the tears come streaming down my face.

GEORGE B. KISTIAKOWSKY

REMINISCENCES OF WARTIME LOS ALAMOS

To refresh my memory about Los Alamos I read a couple of books on the wartime period but got confused and irritated because they certainly didn't jibe with my recollections. Everything in books looks so simple, so easy, and everybody was friends with everybody. But even more annoying was that none of my young associates, who did most of the work, were even mentioned. So I got in touch with people who were there and who are now in the vicinity of Boston, and did some telephoning to my former associates in other cities. In a way, what follows is a group recollection, but I take full responsibility for it.

Before coming to Los Alamos I was probably the senior expert in the whole explosives division of the National Defense Research Committee (NDRC). I began research on explosives in June 1940 (when NDRC was organized) and by 1943 thought I knew something about them. I was also involved in a National Academy of Sciences committee late in 1941, which advised President Roosevelt that the atom bomb was feasible.

The man who deserves full credit for developing the concept of implosion, necessary to explode a plutonium weapon, is S. Neddermeyer. He and his assistant visited our NDRC Explosives Research Laboratory in Bruceton, near Pittsburgh, in the summer of 1943. We made the first implosion charges for them, fired them off, and the visitors went away rather pleased with themselves and with us. The reason that I went to Los Alamos was that James B. Conant, who was chairman of NDRC and knew my views on military high explosives (that they could be made into precision instruments, a view which was very different from that of military ordnance), was also the effective policy maker of the Manhattan District, as the co-chairman of the so-called military policy committee guiding General Groves. I began going to Los Alamos as a consultant in the Fall of 1943, and then pressure was put on me by Oppenheimer and General Groves and particularly Conant, which really mattered, to go there on full time. I didn't want to, partly because I didn't think the bomb would be ready in time and I was interested in helping to win the war. I also had what looked like an awfully interesting overseas assignment all fixed up for myself. Well, instead, unwillingly, I went to Los Alamos. That gave me a wonderful opportunity to act

49

L. Badash, J. O. Hirschfelder and H. P. Broida (eds.), Reminiscences of Los Alamos
1943–1945, 49–65.

as a reluctant bride throughout the life of the project, which helped at times.

After going there on full time at the end of January 1944, I brought there from the explosives division of the NDRC quite a few people, who became the key operators in the implosion project: Linschitz, Koski, Kauzmann, Hornig, Jackson, Patapoff and others, later Roy. I found a very small operation under Neddermeyer — only a few people working — and a continuing angry conflict between Neddermeyer, who is now a professor at the University of Washington, and his boss, Captain Parsons, later Admiral Parsons, now unfortunately dead, who was the head of the Ordnance Division. I was put in the middle, between them by Oppenheimer, and that was a very uncomfortable place. So uncomfortable that two or three months later I wrote a memo to Oppenheimer, asking to be released from the project. That didn't work and instead I got a more authoritative job.

By the summer of 1944 I was supposedly the boss of all the implosion work, but, as I was still under Parsons, the conflicts continued. Theoretically, the official channel of communications was from Neddermeyer to me, to Parsons, to Oppenheimer. Well, it didn't work that way. Basically, Neddermeyer believed that the implosion research should be done by a small group, in a consecutive set of experiments until the right way of doing it was achieved. Now, I don't know whether you know what implosion is. We ended with a spherical charge of high explosives, almost five feet in diameter with a metallic pit in the middle. In the center of the pit was the plutonium fissionable material. Our job was to induce the pit and the plutonium to be compressed in an orderly fashion under the extreme pressure of a detonation wave, many millions of pounds per square inch, into something very much smaller than it normally was, whereupon it would become supercritical. A nuclear reaction would then spread and a big bang follow. Neddermeyer believed that this had to be discovered in a scientific, orderly fashion. Captain Parsons was a Navy ordnance officer, accustomed to developing mass products, and therefore felt very differently about how the work should be done. Soon after I arrived Captain Parsons brought to Los Alamos a Mr. Busby, an old Navy ordnance civilian to be in charge of explosives manufacture. And so the issue very soon became who, Busby or Kisty, [i.e., Kistiakowsky] knew more about explosives. Busby was a little difficult because when you disagreed with him about what was safe and what was unsafe, he would say "and have you ever picked up a man on a shovel?" So Mr. Busby designed and had built the first explosives casting plant. It was a monstrosity from our point of view. Actually, what he had built in the spring of 1944 was never used afterwards.

Then came the summer of 1944. I insisted that we build another plant according to our concepts and proposed a completely new site for it, in Pajarito Canyon. Captain Parsons rejected that and insisted that the existing S-site be expanded and that's where the plant went. But it was built according to our designs, and it worked. One of the reasons I didn't like the S-site was that to get the raw explosives there they had to be trucked all the way over the mesa, right through the center of the Los Alamos project, with the whole theoretical division sitting in offices on one side of the road, Oppenheimer's office (and mine) on the other, and with hundreds of wild WAACs and GIs driving trucks and jeeps there. I can assure you that a truck loaded with five tons of high explosive (H.E.) going off there would have wiped out 90% of the brains in those temporary buildings. The roads leading from the main project to the S-site, which was a few miles south of there, were dangerously rough. Once, so an apocryphal story goes, when General Groves visited Los Alamos, I took him to the S-site in my jeep that had the springs made inoperative by wooden blocks inserted under them. General Groves was rather rotund in shape. As a result of that trip the roads over which H.E. were moved were improved.

We had developed in NDRC complete different notions from those of the Ordnance Corps about what was dangerous and what was safe in handling high explosives. At Los Alamos we handled them by the ton, whereas one gram of explosive going off in your hand will finish off the whole hand. By the end of the war we had cast and machined, following our rules of safety, tens of thousands of H.E. castings without a single accident. And we did it without barricades, which are required by ordnance rules. There was no time to build barricades. So we just worked. The S-site was managed by the Army Captain Jerry Ackerman, a civil engineer before the war, with a few very young Navy lieutenants and ensigns, like Hopper, Wilder, Chapell and a large number of GIs, who were called SEDs (Special Engineer Detachment), doing most of the work. Also there were quite a few civilians such as Price and Gurinsky, who were mostly in a development section, to find how to cast these explosive slurries. They were mixtures of explosives, a so-called Composition B and another explosive, Baratol, which we had developed at the Bruceton (Pittsburgh) NDRC Explosives Laboratory, and which was used for lenses. Up to 25 tons of H.E. was trucked up the hill monthly during the most active period. The manufacture of the H.E. charges was really hard, tough and dirty work. A lot of people thought it was also very dangerous and because of that I went to S-site very frequently and tried out new operations simply to show confidence in what we were preaching.

One of the very important contributors to our success was Master Sergeant

Tenney, in peacetime a doctor of physics, who developed non-destructive X-ray inspection of castings. Until the middle of 1944 all we could do was to cut a casting with a saw to see whether it was homogeneous or had bubbles and cracks. Gradually Jerry Tenney's project grew into a mammoth operation with million volt X-ray and gamma ray sources to inspect all the castings.

The implosion experimentation involved exploding the castings and trying to find out whether they did what we hoped they would do to the metal inside. It was originally carried on by Neddermeyer, Greisen, Lynch, Kauffman, and an increasing number of SED assistants. What became more and more clear was that if an explosive charge was detonated simultaneously from several points then trouble developed. Remember that detonation waves are like acoustic waves, more or less, but they develop enormous pressures and travel through the explosive at a speed of seven to eight thousand meters per second (five miles per second). At the point where two detonation waves meet, a metal core is squeezed into a high velocity jet and complete chaos develops. In other words, this sort of thing makes implosion impossible. To solve that trouble, first Linschitz, then Koski, then Jim Tuck, a member of the British technical mission who was sharing in our work, began experiments with explosive *lenses*. I won't go into details except to say that the principle on which these lenses is based is just like that of the optical lenses, where you have two media of different refractive index, such as air and glass, and therefore a different velocity of light in each. If you have two explosives with different detonation velocities, and you put them together in the right way you can shape the wave, and instead of having it expand, make it to converge. It's not simple, but it can be done. And I arranged with the Bruceton NDRC laboratory, particularly with Drs. MacDougall and Eyster, to develop an explosive, Baratol, for the slower component of the lens.

There is an interesting story about Baratol. Lord Cherwell, who was originally professor of physics Lindemann at Oxford, and was *the* science advisor to Winston Churchill, came to Los Alamos in 1944. He had been a very kind and helpful host when I spent part of the spring and summer of 1941 in England learning about explosives technology from the British. He was very considerate then and took me to his 'private' explosives research laboratories. His staff there was developing trick things for the British commandos. Very interesting ideas. Well, so he fancied himself as an explosives expert. I was trying to be extra polite, so I took him around Los Alamos and gave him absolutely the whole story. He listened to me and told me that Baratol was going to be no good for the lenses, that I should use commercial dynamite. I explained to him why, on theoretical grounds, dynamite couldn't work and we parted friends. A little later Oppenheimer called me in to tell me about Churchill's personal cable to Roosevelt saying that certain people, specifically Kistiakowsky at Los Alamos, are barking up the wrong tree, since Baratol is not going to work, and that they should use dynamite. I suggested to Oppenheimer that he could tell the sender to go to hell. Naturally it ended by my agreeing to set up a group to study dynamite lenses. So I went through a rather large X (Explosives) Division personnel list, singled out individuals who hadn't contributed anything, constituted a group out of them, and so dynamite didn't delay the project in the slightest.

Because of organizational difficulties and the growing importance of implosion, in July and August 1944 there was a big reorganization. Parsons continued to head the Ordnance Division but two new divisions were created, X and G. I ran the X, which did the high explosive charges development work and the experiments with the already developed observational methods on implosion dynamics, also lens development, and some other tasks. We ran this division in a very democratic fashion, with frequent meetings for discussing the results of what we had done, scheduling new experiments, scheduling and allocating the output from the S-site and so on. Unfortunately, my young associates like Linschitz, Koski, Kauzmann, Tenney, and others were regarded, high above, as too young to be made into group leaders because then they would be on a par with people like Edwin McMillan, Luis Alvarez, and Edward Teller. So they were made into section leaders, and Commander Norris Bradbury, professor of physics at Stanford, was brought in August from the Dahlgren Naval Proving Grounds to head group X-1, which was comprised of about five sections headed by these youngsters. They operated the sections in a pretty autonomous fashion because each had his own experimental site for safety reasons — maybe as much as a mile apart. And the sites were fairly self-contained, perhaps even possessing a lathe, a few draftsmen and so on.

The G (Gadget Physics) Division, Robert Bacher being put in charge of it, undertook to develop new experimental methods of observing the movements of the pit after being struck by the detonation wave, and also to design the pit, including of course the plutonium sphere. I might add that this reorganization was partly due to the theoretical conclusion that implosion was the only way in which an effective plutonium bomb could be made. I welcomed this separation of functions because by then the total implosion staff became much too difficult to manage. But in a few months' time our relations with the G Division became not very happy because the G groups put tremendous demands on the S-site to provide them with tricky explosives castings, which the S-site simply could not meet without stopping supplies for the X Division experiments and interfering with its own development of better casting methods. So we as a group began to feel that the G Division was merely delaying the progress of the implosion project by working on such complex experimental methods that they would never yield anything useful in time. I must admit that G Division people didn't share our views. Since Bob Bacher is not a shrinking violet, he and I had some very heated arguments. Oppenheimer, I think, backed him more than me — maybe Bob Bacher has a different view — and so Oppie even brought Charles Lauritsen from Cal Tech to Los Alamos to help the X Division as his, Oppenheimer's, personal assistant. That added to my troubles, because the key source of delay in providing more castings was that we couldn't get enough molds in which to make castings. These molds were really precision devices because the shapes into which explosives had to be cast were such that we could not machine rough shaped castings with the tools available to us at the S-site. They had to be cast into precise shapes. Some were not small. For instance, full sized castings weighed up to 100 pounds or so. Some X Division people, especially Bob Henderson and Earl Long, were of tremendous help in providing precision molds. Gradually huge machine shops were created at Los Alamos for this and other purposes — more than 500 machinists and toolmakers were working there at one time.

I think that certainly not the least factor in our success was Master Sergeant Fitzpatrick (SED), our X Division procurement and scrounging wizard. We learned early in the game that if we followed the rules of the Manhattan District, we wouldn't get anywhere. The slow facility construction and procurement delays, and the shortages of various materials were really a painful problem. When Charlie Lauritsen came to Los Alamos he was critical of our procurement efforts. So Charlie said that his Cal Tech Navy rocket project staff will solve our molds problem; but they didn't in time, and so the spring of 1945 was very short on molds and very rich in recriminations.

By late 1944 so much pessimism was developing about our ability to build satisfactory lenses that Captain Parsons began urging (and he was not alone in this) that we give up lenses completely and try somehow to patch up the non-lens type of implosion. So, in early 1945 we had a top-level meeting with General Groves present in which a kind of battle royal was fought, in a friendly way, between Parsons and me because I felt that we couldn't patch implosion without lenses and he felt we couldn't make the lenses. Oppenheimer in the end decided for the lenses and that was that.

In the early days of the implosion project, to get detonation started all over a sphere of explosives there was one electric detonator firing a branching circuit of Primacord. Primacord is a rope-like detonating fuse which transmits the detonation wave at a speed of some three miles per second. The ends of this circuit were embedded at appropriate points in the explosive charge. The use of multiple electric detonators, or blasting caps, was objected to by Parsons' Ordnance Division because the Army-Navy rules require that there be a mechanical 'gate,' that is a piece of metal, between a detonator and the main explosive charge which has to be withdrawn before firing. For instance, in an artillary shell, the 'gate' is withdrawn by centrifugal force after the spinning shell leaves the muzzle of the gun. This makes the shell safer to handle before firing, because detonators are dangerously sensitive to impact and heat. To install as many such gates as would be necessary in the implosion device was just an engineering nightmare. Furthermore, the electric detonators then in existence had horribly poor timing, so that to explode them simultaneously looked, to us at least, absolutely impossible. Mind you, since detonation waves travel almost a centimeter per micro-second, the timing we were concerned about were fractions of a millionth of a second. We did a lot of experimenting with Primacord ordered in fancy, expensive, special batches. Lt. Shafer did most of this work. The Primacord just wasn't good enough.

Luis Alvarez in the summer of 1944 started experimenting with electric detonators and, I must say, completely to my surprise, found a way of setting them off in such a way that the simultaneity was very acceptable. Then new detonators were so designed by Alvarez, with my help, that we were able to persuade Captain Parsons to drop the requirement of 'mechanical gates.' So we abandoned Primacord. Then Bainbridge, Fussell, and Hornig in our division began work on what we called the X unit, an electric device to fire these new detonators simultaneously. Alvarez moved to other problems in the G Division, Bainbridge went to head the Alamogordo or Trinity test site, and Greisen in our division took over the detonators in the spring of 1945. That turned out to be a very nasty and unfinished problem because we couldn't get reproducibility with the Alvarez detonators. Most of them worked fine, but since one had to have many for each implosion, and even the failure of one could be a catastrophic failure, even doubling the circuit didn't give adequate assurance. So Kauzmann and Jackson did a very clever piece of physico-chemical research on the explosive charge in the detonators and found a way of making them reproducible.

To test or not to test the plutonium bomb was a very hot issue in the fall of 1944. Oppenheimer and I were pleading with General Groves that there had to be a test because the whole scheme was so uncertain. But General

Groves said he couldn't afford to lose all that plutonium if the chemical explosive went off but there were no nuclear explosion. General Groves was very sensitive about what would happen to him after the war and whenever he didn't like something, he'd say, "Think of me standing before a U. S. Senate committee after the war when it asks me: 'General Groves, why did you spend *this* million or *that* million of dollars?' " Well, it was difficult to answer that, too. So we proposed to test the bomb in a confining vessel, and a couple of very bright engineers in our division, Carlson and Henderson, designed Jumbo, a 200 ton ellipsoidal steel tank, with wall thickness nearly twelve inches, in the center of which, they believed, several tons of high explosive could be exploded and, if there were no nuclear reaction, every-thing would stay inside, although the vessel would stretch out quite a bit. And so plutonium could be recovered after everything cooled down. In the spring of 1945, Jumbo was delivered to the Trinity test site, but by then more plutonium was coming off the production line and we felt much more confident of implosion. So Jumbo was set up just half a mile away from the Trinity tower, and never used.

By the way, a sad story about Jumbo — Groves was very sensitive about Jumbo and kept accusing me personally about it. Once he said to me, "Now you are responsible for Jumbo, and it is not being used, so what am I going to tell to the Senate? Look here, we got that cyclotron from Harvard for Los Alamos. How about you convincing Conant that when he goes back to the presidency of Harvard he should accept Jumbo in trade for the cyclotron at Los Alamos?" Actually Jumbo had a very sad end after the war. In the fall of 1945 Groves suddenly remembered that ghastly thing and told his aide-de-camp to arrange for a test, so at least he would be able to tell the Senate committee that the thing had been used. Well, you know the way it is in the Army. The orders filtered down, down from Washington to Albuquerque, Albuquerque to White Sands Proving Grounds, and so on. It ended with some poor lieutenant taking a box of high explosive and instead of suspending it in the center of Jumbo, putting it on the bottom of Jumbo and firing it. Naturally, the high explosive knocked out a nice clean hole in Jumbo, so that was the end of the matter. The U.S. Senate never caught on to the Jumbo extravaganza and its battered remains are still there, half buried near the Trinity site.

The Special Engineering Detachment (SED) was very important. General Groves insisted that, in distinction to technical staff civilians, most of whom lived quite comfortably in a highly hierarchically organized society, the enlisted personnel be given only what the Army regulations stipulated as the minimum comforts: minimum housing, minimum recreation, minimum food facilities. And this meant 40 square feet per man in the barracks, including part of the recreation area. Try to recreate yourself in that area. So the poor SEDs, of whom we had more than a thousand at the end of the war, really felt themselves the pariahs of Los Alamos. And moreover the commanding officer of the SED detachment was a South Bostonian. You must have read a little about South Boston in the Fall of 1974 when the Boston schools opened. Well, he was a true South Bostonian and besides he had been wounded in the heel, in fact the back of the heel, on the first day of Eisenhower's African landing. As a result, he simply hated the world, and especially the longhaired scientists, notably those from Harvard. Since he was not told, as many other military weren't (nor the machinists, of course), what the purpose of Los Alamos was, he loudly described all of us as draft dodgers who were just escaping Army service and having fun here. He insisted that the SEDs be awakened by a reveille and be mustered daily and do calisthenics and keep the barracks in order and even wear caps and salute officers on the streets. These were insulting ideas to most of the SEDs. Since my division had the most SEDs and we had very good relations, there finally came to me an SED delegation who said they were going to complain to the War Department unless that officer was removed. I said, "look, this is mutiny in war time – don't," but they said "then do something about it." So I naturally went to Oppenheimer and he, not for the first time, argued with General Groves – to absolutely no avail. Later, when Groves came again to Los Alamos I asked permission to talk to him. He said, "yes, while I am being driven back to Albuquerque." So after midnight I got into his car and we went on that 2½ hour trip to Albuquerque and argued about SED. Of course Groves immediately told me that as a civilian I had no business to tell him anything about Army matters. And I said that the SEDs were part of my technical staff, they had to report to me, they had to work for me, and therefore I had the authority. Well, I got absolutely nowhere. I then used my ultimate weapon: I said I would resign. Still seemingly no effect. Well, I didn't have time to resign, because in a little while the South Bostonian became the Manager of the Officer's Mess and the SED got a new commanding officer, a very nice tall Texan, who made absolutely no disciplinary demands and spent most of his time drinking in the company of a very cute WAAC. Eventually both slid with their jeep down a canyon-side, but that was after the war. And they didn't get hurt, they were so relaxed.

The Trinity test was the climax of our work. The site is 200 miles south of Los Alamos in a part of a bombing range called Jornada del Muerto desert, not very near Alamogordo, but in that region. The G Division and Oppenheimer insisted that a perfect replica of the charge to be exploded at the

Trinity test be fired at Los Alamos using the so-called magnetic method observation technique. We were desperately short of adequate H.E. castings. A few
days, maybe a week, before the test, after borrowing an electric dentist's
drill, I spent most of a night drilling holes in imperfect castings where Sergeant
Tenney had discovered bubbles of air. I reached the bubbles and then poured
molten explosives into these channels to fill the bubbles and drill holes and in
that way make the castings better, because we knew that when there were
bubbles the detonation wave became distorted. Just like the bubbles in an
optical lens, or scratches do. In that way we repaired enough castings so that
the two assemblies, each weighing several tons, were assembled, with Norris
Bradbury in charge of this particular job.

The assembly to be fired at Trinity had to be trucked through Santa Fe
and Albuquerque and a lot of people outside the X Division thought that
these assemblies were far more dangerous than ordinary iron aircraft bombs,
which they really weren't at all. So I rather vividly remember that I got on a
truck with a loaded assembly and drove it around the Los Alamos roads
which were certainly worse than any roads we would encounter on the trip to
the Alamogordo site, just to show that nothing would happen. A few minutes
past midnight on Friday the 13th, my choice, because I believed in unorthodox luck, Bradbury and I took the assembly in a convoy to the Trinity site.

Bainbridge had been in charge of planning and then developing the Trinity
test site since 1944. But in 1945, as this project expanded and became more
and more tense, he was separated from the X Division and became the head
of a new TR Division. This grew of course to a very large size, because so
many experiments and observations were being readied for and at the Trinity
site. Hornig, about July 10, 1945, (six days before actual test), took several
X units to Trinity because various observational instruments had to be
triggered by the firing of the X unit for simultaneity. And so they had to use
the X unit to assure synchronization. A day or two later there came a violent
storm and the X unit being used fired prematurely by itself. Well, a human
storm followed over Hornig's head for incompetent design because people
imagined this happening when it was connected to the bomb. Until, that is,
Hornig discovered that the grounding wire was pulled off or busted or something and the X unit got a huge static charge from the lightning. That sort of
relieved the human storm.

We arrived two days later, early on Friday, to encounter another emotional
scene. The second X unit failed dismally the evening before and Don Hornig
spent most of the night being grilled — and the word is grilled, not questioned
— by Oppenheimer and Bacher, being accused of incompetent work and so

on. As soon as we arrived I was told what they thought of me as a manager of these incompetent youngsters. So Don Hornig and I went finally to look at the X unit which was located under the bomb tower and discovered that the people who were using it while testing their instruments, horribly abused the unit. The unit was designed to be used only once, and it was tested about ten times to make sure that it worked fine. Well, these characters used it in rapid succession hundreds of times and overheated it so that some soldered joints melted. This discovery relaxed the atmosphere in the headquarters. But Saturday morning another awful thing happened. A telephone message came from Los Alamos that the G Division's magnetic method group found their H.E. assembly's detonation so bad that they could guarantee the failure of the bomb at Trinity. So of course I immediately became the chief villain and everybody lectured me — Oppenheimer, Groves, Bush, who was there by then with Conant. Did they tell me what was wrong with me! Only Conant was reasonable. Sunday morning another phone call came with wonderful news. Hans Bethe spent the whole night of Saturday analyzing the electromagnetic theory of this experiment and discovered that the instrumental design was such that even a perfect implosion could not have produced oscilloscope records different from what was observed. So I became again acceptable to local high society.

On Friday or Saturday, Bacher and his group inserted the plutonium into the pit, then Bradbury and a couple of SEDs replaced the H.E. castings, which had been taken out to be able to get into the pit. The bomb was hoisted to the top of the hundred foot tower. There Hornig installed a fresh X unit and Linschitz and Kalecka, an SED in our division, finally inserted the detonators into the charge. I had very little to do the last two days, just watch others. Sunday night I spent partly up on the bomb tower with Bainbridge and two others below me because weather was bad for the test which therefore had to be delayed and Groves insisted that there was danger of sabotage to the bomb — a perfectly idiotic idea. So we were supposed to watch it with a sub-machine gun in the hands of Captain Bush and that sort of thing. Finally, the decision to fire was made at five-thirty in the morning. We drove back, first unlocking the box containing the safety switch and all of us solemnly watching as Bainbridge closed the switch and locked the box. Then we drove a mile, repeated the operation on a switch box in a trench, and finally got to the locked box in the control bunker six miles away, opened it and closed that switch. The thing was ready to be fired. I had nothing to do and so just before the time counting came to zero I went up to the top of the control bunker, put on dark glasses and turned away from the tower. I didn't think anything would happen to me, although I was sure that implosion would work, because I was rather convinced that the physicists exaggerated what would happen from a nuclear point of view. Well, I was as wrong as they were on Saturday. As soon as the explosion took place, Oppenheimer and others rushed to join me and I slapped Oppenheimer on the back and said, "Oppie, you owe me ten dollars" because in that desperate period when I was being accused as the world's worst villain, who would be forever damned by the physicists for failing the project, I said to Oppenheimer, "I bet you my whole month's salary against ten dollars that implosion will work." I still have that bill, with Oppenheimer's signature. Now, there is something more to that story, because after VJ day there was a lot of nonsense published in Sante Fe and other more or less local papers about the Trinity test, that I was, as they put it, a temperamental Russian who lost his self-control and embraced and kissed Oppenheimer, whereas all I did was slap him on the back and say, "Oppie, you owe me ten bucks." So, after VJ day, when we had a sort of post-mortem, and all the group leaders of the project must have been there, Oppie said, "you must have read these stories; I want to testify that they are wrong, George never kissed me, but I am now going to kiss him." He did that and gave me the ten dollar bill.

Let me now make a few remarks of a more personal nature. General Groves was really a terror to his subordinates. He was a skillful manager, but he always did things so as to make life difficult for the subordinates. However, he and I had perfectly good relations. I think he saw me as more manly than the effete physicists because of my explosives work. I was to him a sort of kindred spirit.

I had very good relations also with Oppenheimer. He had an incredible ability to have all the threads of this enormous project in his mind and to make the right technical decisions. As to his managerial skills, that's a little different story. Conant regularly came to the project because he was the really active member of the Military Policy Committee, and he and I usually had private technical conversations. I met several times with Niels Bohr when he was at Los Alamos but only because he wanted to know from me how implosion was progressing. I naturally dealt a lot with the British group – Chadwick, Peierls, Tuck, Penney. The man who eventually turned out to be a spy, Fuchs, managed to put himself into a very important position. He was a member of the Theoretical Physics Division, like Professor Hirschfelder, but he also arranged to be appointed as the liaison between Theoretical Physics and X Division, so that he sat in on all our discussions and planning meetings.

At Los Alamos I was treated like a VIP. I had special housing, a tiny little stone cabin that used to be a diesel station before the war. Oppenheimer sold me, for a nominal sum, one of his saddle horses, a beautiful Quarter horse named Crisis. The Army maintained a horse stable in Los Alamos and for a small fee took care of a half-dozen or so private horses. We never worked on Sundays, that was a hard and fast rule, so on Sundays I rode horseback in the mountains. In the winter of 1944–1945 we built a ski slope nearby using explosives to cut down the trees. We had a lot of surplus plastic explosive, the demolition explosive, and if one builds a half necklace around the tree, then the explosion cuts it as if you had a chain saw – and it's faster. A little noisier, though. Then we scrounged equipment to build a rope tow and it became a nice little ski slope. From my friends in Washington, I got a couple of skimobiles which weren't like the modern skimobiles but more like jeeps on tracks. They weren't very good but we managed to use them to go skiing further out. I played a lot of poker with important people like Johnny Von Neumann, Stan Ulam, etc. You see, before coming to Los Alamos, I went through a very rigorous and expensive poker training school in Washington, headed by Roger Adams, NDRC member and my boss. So when I came to Los Alamos I discovered that these people didn't know how to play poker and offered to teach them. At the end of the evening they got annoyed occasionally when we added up the chips. I used to point out that if they had tried to learn violin playing, it would cost them even more per hour. Unfortunately, before the end of the war, these great theoretical minds caught onto poker and the evening's accounts became less attractive from my point of view.

We did quite a lot of partying on Saturday nights, but there was *one big* party after VJ day in Bacher's house, and by the time we had quite a few refreshments my physicist friends started telling me that I must arrange for a 21- gun salute. Of course I didn't have the guns. Finally, I got into my jeep and got one of my younger friends out of bed, who then insisted on driving the jeep. We went to the H.E. magazines, got out 21 boxes of Composition B (50 pounds each), laid them out on the field (I was stumbling a little because the field was very rough) and fired them off. It was a very impressive performance but when I got back to the party those bastards told me I fired 22 shots.

DISCUSSION

Question: What field did Von Neumann work in in those days?
Kistiakowsky: Von Neumann throughout the war was a consultant to Los Alamos. He spent, toward the end, I think, as much as half of his time there.

When he was there he was a member of the Theoretical Physics Division. But even earlier, you see, he did the important job of convincing Oppenheimer and others who questioned Neddermeyer that if the implosion were perfect, the forces acting on the pit would be such as to compress the plutonium to a density that one would presumably find near the center of a star or something like that, and thereby very rapidly change it from a sub-critical, in the nuclear sense, to a highly super-critical assembly. Von Neumann was very much in favor of implosion. In 1941–1942 he did some important theoretical work for the explosives division which I headed in NDRC (Division 8), so his interest in explosives was genuine.

Question: Some of the information passed by the Rosenbergs to the Russians concerned the lens and your aspect of the project. How critical was that information?
Kistiakowsky: That is something I have been asked about before. From reading Mr. Khrushchev's memoirs and some other information, I gather that the Russians had started the atom bomb project quite early in the war. For instance, there is the story that when Stalin met with Truman in July of 1945 and Truman told him of a successful Trinity test the day before, Stalin acted as if it were of no interest to him at all. According to Khrushchev, however, he immediately sent a message to Moscow to put Beria in charge of their project. So the project must have been going. Under those conditions the very crude sketches of Greenglass could not have been of great importance. Maybe of no importance whatever. I think what probably was far more important is what Fuchs transmitted because he was able to send detailed results of calculations and also the information on problems of timing the detonators and a lot of other things like that. I think though that even this did not make a tremendous difference, because it turned out that to cause the implosion is a much easier job than we thought it was. But it might have accelerated the Russian bomb by a year or two. This would be my guess. But it's only a guess.

Question: I have often wondered how much sooner World War II would have been over if there had been no atom bomb project and so many of our scientists and resources had not been shifted from conventional arms to the atom bomb. Surely this would have speeded up the German part of the war and probably the Japanese.
Kistiakowsky: Well, it's a speculation in which I will not engage. After all, the Manhattan District, which was active during the war, was not a very large project – two billion dollars in three years – at a time when the total war

effort was costing nearly a hundred billion dollars a year. So when you refer to resources, you mean the very special resources that were in tremendous demand, but whether they would have altered the war's progress, I don't know. Whether Japan would have delayed surrendering is also an unsolved question. I think that what one now knows suggests that they would have surrendered anyway. We were told, and I am not sure whether the military intelligence people came to Los Alamos or whether Oppenheimer brought the news from Washington, that according to our Naval Intelligence, (as you know there are several kinds of intelligence: there is human intelligence, there's animal intelligence, there's the military intelligence), that according to Naval Intelligence Japan was ready to continue fighting the war and that we would have to go through with our plan of invading the main Japanese islands in November 1945 and that there would probably be a million casualties before the war ended. That information had a trememdous impact on me, on my thinking about the military use of the bomb. Since then, it's become pretty clear that Japan was really far closer politically to surrender than that estimate indicated.

Question: Do you think that it is easy for a small group of terrorists to construct nuclear weapons?
Kistiakowsky: That is an estimate which I've heard advanced by Dr. Theodore Taylor. I don't think so. Perhaps I am too proud of our work. I think you couldn't build a reliable bomb without doing some experimentation with explosives and that experimentation is more than a small private group could undertake. On the other hand, it would not require the resources of a Soviet Union or India to build a bomb. It could be done on very much of a shoestring. The bomb wouldn't perform as well, but there is quite a leeway. Certainly anything approaching the complexity of the Trinity bomb and of the implosion bomb dropped over Japan, I would say, would require a very major effort. Assuming you already have the fissionable material, I would estimate several million dollars worth of effort and a couple years. One could build a much cruder bomb, but even then it would require some resources. Let's say a country possessing one technical university and some military forces and some kind of proving grounds could do it.

Question: What does a lens have to do with implosion? Is it a lens like the lens in my glasses?
Kistiakowsky: It's not an optical lens, though it functions similarly. The explosive lens bends the explosion wave going through the explosive. To an

innocent it looks like another piece of explosive. It's made out of two different explosives, and that's the secret of it.

Question: Were the scientists at all disturbed that Truman decided to drop the bomb on a population center instead of testing the bomb in an unpopulated area? Did the scientists have anything to say about how the bomb would be used?

Kistiakowsky: Well, the scientists in the Chicago part of the Manhattan District — the Met Lab, which by the spring of 1945 was almost inactive because it had done its job — became extremely active trying to stop the military use of the bomb over a city, urging a harmless demonstration instead. At Los Alamos we had some conversations on the subject and I must admit that my own position was that the atom bomb is no worse than the fire raids which our B-29s were doing daily in Japan, and anything to end the war quickly was the thing to do. Other people felt differently but there was no organized movement at Los Alamos to stop the bomb use. I changed my mind afterwards but I was very much influenced by the military intelligence estimate of what would happen that summer. I won't go into detail of some of the more technical arguments against making a demonstration, but it looked unfeasible to me.

Question: Was the second bomb that was dropped on Nagasaki necessary? Didn't just a very minor point in the surrender terms remain, namely the safety of the Emperor?

Kistiakowsky: The Japanese position was that they could not surrender unconditionally because that would probably mean that the Emperor would be dethroned and maybe even executed. They would be willing to surrender if the monarchy would be retained. I think our excuse for dropping the second bomb was very weak. As I recall it, the argument was: let's drop them quickly — one, two — to give the Japanese government the idea that we have an unlimited supply. We only had two. It would have been at least a month, or maybe two, before another bomb could be produced and dropped over Japan. But, of course, we at Los Alamos had absolutely no control over their use once the bombs were shipped overseas. This happened within ten days or so after Trinity, as soon as we manufactured the new set of H.E. castings. The other type of bomb, the gun type, which was used over Hiroshima, was shipped overseas even before the Trinity test. After that, there was nothing we could change.

Question: You said that you had to cut through red tape, to cut corners in order to make any progress. I wonder if you would describe some of those methods.

Kistiakowsky: No sir, they will remain secret. Well, let me tell you this much. There was an extraordinarily elaborate procurement system for military security purposes. We had to order everything from an office in Los Angeles. That office, acting as a part of the University of California, and not of Los Alamos, which was secret, then ordered things from all over the country. These things were shipped to Los Angeles and then were delivered from there to Los Alamos. That resulted in terrible delays and errors. We found ways of getting around them.

Question: Was there any awareness in your group of German progress along the same lines?

Kistiakowsky: The rumour of German progress was the thing that was egging us on. But by the end of 1944 it became pretty obvious that the Germans didn't have the bomb and wouldn't have it in time. Then the argument presented to us became that we must end the war with Japan as quickly as possible. But in earlier days it was a very real fear that the German bomb would be built and would win the war for them. Aa a matter of fact, that interesting job assignment abroad I missed because of going to Los Alamos had to do with trying to find out what the Nazis were doing on the A-bomb.

Question: Was there just one piece of plutonium in the bomb?

Kistiakowsky: We had plutonium cast into two hemispheres that fitted very neatly together. And we had no more plutonium at Los Alamos. The plutonium, however, was being manufactured very rapidly at Hanford and so within a relatively short time after Trinity a second sphere was made at Los Alamos and taken to Tinian Island.

Question: What effect did the ending of the war with Germany have on Los Alamos?

Kistiakowsky: We celebrated it.

Question: I mean in terms of your feelings about the project, its motivation, things like that.

Kistiakowsky: Well, naturally it became less important, the whole thing became less intense, as we began to realize that we were certain to win the war. But, as I said, the feeling was conveyed to us that Japan was very far from surrender, that the war would continue for a long time. That continued to provide, I think, the emotional cohesion of the laboratory staff.

JOSEPH O. HIRSCHFELDER

THE SCIENTIFIC AND TECHNOLOGICAL MIRACLE AT LOS ALAMOS

I am concerned with the ways and means by which science can help to make this a better world in which to live. This is an age of wondrous discoveries: we can now cure many kinds of diseases, increase our production of food, produce large amounts of energy, and make life much easier and pleasanter. However, these same discoveries are leading to an explosion in our population, the depletion of our natural resources, and the poisoning of our atmosphere and water with pollution. As Pope Paul VI said in his 1966 speech to the Pontifical Academy of Science, "The discoveries of science are indeed wondrous, but unless these discoveries are used for the good of mankind and to make this a better world, then they are worthless." Or, as Isaac Asimov said in an after dinner speech at an American Chemical Society meeting, "Any fool can make a great discovery, but it takes a genius to figure out the consequences." Thus, as a result of public apathy, greed, and ignorance, many of the scientific discoveries have been technologically misused and the world-wide situation which we face today is indeed a mess. The future for our children and their children looks very bleak — that is, unless we can produce a set of new scientific-technological miracles which are needed to solve our present problems.

I believe in scientific-technological miracles since I saw one performed at Los Alamos during World War II. The very best scientists and engineers were enlisted in the Manhattan Project. They were given overriding priorities. They got everything which they deemed essential to their program; the cost was unimportant. They had the full cooperation of everyone and they, themselves, devoted long hours in mixing together their ingenuity and technical skills. They lived in an isolated village high up in the Jemez Mountains of New Mexico where they worked together, played together, and lived together free of all outside distractions. In a period of two-and-a-half years, they produced the miracle — an atomic bomb which creates temperatures of the order of $50\,000\,000°C$ (or 15 000 times as hot as molten iron), and pressures of the order of 20 000 000 atmospheres (or pressure greater than at the center of the earth), while unleashing the tremendous energy stored in the atomic nuclei. Actually, by taking all of these scientists and engineers off conventional military research on guns and rockets, it probably took an extra six

67

L. Badash, J. O. Hirschfelder and H. P. Broida (eds.), Reminiscences of Los Alamos
1943–1945, 67–88.

months before the peace with Germany was achieved. I do not know what the effect would have been on the Japanese war. When a large number of top scientists and engineers and a large fraction of resources are devoted to one purpose, then the other things are bound to suffer.

The atom bomb and the research on it was, however, a necessary safety factor, because we had every reason to believe that at least Germany was working on it and working on it hard. The recruiting slogan of the Manhattan Project was "Help win the war to end all wars", and we all felt that this was true. We thought that once World War II had been finished, wars would be much too horrible — there could never be another war, particularly if fearsome atomic energy was made available.

At Los Alamos during World War II there was no moral issue with respect to working on the atom bomb. Everyone was agreed on the necessity of stopping Hitler and the Japanese from destroying the free world. It was not an academic question — our friends and relatives were being killed and we, ourselves, were desperately afraid. Since 1938 when Otto Hahn and Fritz Strassmann discovered nuclear fission and every physicist in the world became aware of the tremendous amount of energy released thereby, the fabrication of atom bombs became inevitable. Although we had no direct intelligence information to confirm it, we were virtually sure that the Germans were making an all-out effort to produce an atom bomb — and very likely the Germans were far ahead of us. Intelligence reports indicated that the Germans were commandeering huge amounts of heavy water from the Norwegian distillation plants; Germany had access to the rich pitchblende (uranium ore) deposits in the Czechoslovakian mountains; Germany's engineering proficiency was second to none. Although a considerable number of German scientists left their homeland rather than work under the Nazi regime, many distinguished physicists, chemists, engineers, etc. did remain there. For example, we had tremendous respect for the genius of Werner Heisenberg who was one of the founders of modern quantum mechanics. It seemed reasonable that Heisenberg might be working on the production of an atom bomb since just prior to the war he had turned his attention to practical problems and developed the theory of turbulence in aero- and hydrodynamics. Then there was Paul Harteck, a very brilliant radiochemist who had been closely associated with the work of Hahn and Strassmann and who would be very clever in designing equipment to separate the uranium isotopes. We had ample reason to respect the ability of Döring, an aerodynamical physicist, to do the actual designing of an atom bomb since he had developed the shape charges which made it possible for the Germans to penetrate the Maginot Line. These shape charges use the same type of focusing of shock waves as we used in the atomic bomb implosions; the idea for the shape charge was the Monroe Effect which was discovered before World War I by an obscure George Washington University physicist who used explosive charges to etch the intricate outlines of leaves and flowers onto steel plates (some of his plates have been put together to form a fireplace screen at the Cosmos Club in Washington). Indeed, it was rather curious that this Monroe Effect was in all of the physics textbooks of the 1910 and 1920 period. If anybody had read those elementary physics texts, they could very well have gotten the idea of the shape charges and World War I might have been entirely different.

And finally, we thoroughly respected the ability of Germany's other ordnance scientists and engineers such as Werner Von Braun who developed the V-2 rocketmissile which had the capability of delivering an atomic bomb for distances of hundreds of miles. Thus, the scientists at Los Alamos felt that they were in a grim race to make the atom bomb. The whole fate of the civilized world depended upon our succeeding before the Germans! We never seriously considered the possibility of the Japanese manufacturing an atom bomb, since Japan did not have the industrial capacity either to make huge diffusion plants, such as at Oak Ridge for concentrating uranium 235 (five stories high, 150 yards wide, and one-third of a mile long!), or to make the large amount of exceedingly pure graphite required to build the plutonium piles at Hanford.[1]

Although most Americans in the 1940's thought of the Japanese as copycats, American physicists knew and respected a considerable number of Japanese physicists. To illustrate the capability of the Japanese physicists, within a day after the Hiroshima bomb had exploded, they had reported to the Emperor that this was indeed an atom bomb, and they had started to measure the induced radioactivity of the sulfur in the insulation of the electric power and telephone lines throughout Hiroshima. As a result they were able to plot the flux of neutrons emitted from the bomb and even the velocity spectrum of these neutrons. Thus, within days, they had discovered a great many of the intimate details of our atom bomb and how it functioned. Throughout World War II we lived with the fear that someone in any enemy country would discover a simple inexpensive method for separating uranium isotopes since, from a theoretical standpoint, this is possible. Imagine our surprise after World War II to find that neither Germany nor Japan had made an all-out effort to construct an atom bomb — the sheer magic of large-scale atomic energy was too fantastic for Hitler to comprehend! A lot of people thought that Roosevelt in the United States was overly optimistic because he grasped at this idea of harnessing tremendous amounts of energy in this manner.

It is an open question as to whether the world is better or worse for our having made the atom bomb. Hiroshima and Nagasaki served as a warning, but the political leaders of the world have still not gotten the message: warfare is no longer a rational means of settling the differences between nations; the fate of the world depends upon the success of arbitration and negotiations. After Otto Hahn's and Fritz Strassmann's discovery it became evident that sooner or later some country would make an atom bomb. If an atom bomb had not been made and detonated in World War II, the world would be unprepared to cope with the tremendous threat of nuclear warfare. The first country to produce an atom bomb could have used nuclear blackmail to conquer the world. Now, at least, we are aware that there is no adequate defense against nuclear warfare. We are rapidly entering a period in which any country can annihilate any other country; moreover, there are other weapons, some biological, for which there is no adequate defense. Thus, warfare is no longer a rational means of settling differences between nations.

On a cold, rainy Sunday night in the winter of 1945, Oppenheimer assembled most of the Los Alamos scientists in a small wooden chapel and, between claps of lightning, explained how all of us would live our lives in fear, but how fear itself might be responsible for maintaining peace until at long last all of the nations had learned to settle their differences in a reasonable manner. Oppenheimer gave us a number of examples of wars in Europe which had extended over a great many years, but because of mutual fear, there had been very little bloodshed.

In order to make an atom bomb, it is first necessary to produce the pure fissionable materials. This was a tremendous task which required a great scientific and engineering effort. Thus, the Manhattan Project set up research laboratories at the Universities of California, Chicago, Columbia, and Iowa State as well as the laboratories and production facilities at Oak Ridge and Hanford. General Groves and the Army Engineers were in charge of the Manhattan Project. However, Vannevar Bush, James Conant, and Richard Tolman were very influential in making the high policy decisions. Thus General Groves' principal function was to expedite policies which were made by other people. He was extremely hard working and efficient, but he had a knack for saying the wrong thing and getting people angry. He never understood the scientists and they thoroughly hated him. For example, when he encouraged his daughter to go into physics, she said to him, "But Father, you know what the physicists think of you!" I think mainly that General Groves' extreme egotism created this impression. He took credit for everything. I remember one time he talked to a group of WAC's and SED's, our military colleagues at Los Alamos, and he got up in the front of the podium and said, "The reason why I am here today is to introduce you to your boss' boss' boss," and that was himself.

The original group of scientists who formed Los Alamos were University of California physicists under the direction of Robert Oppenheimer. Their idea was to make the atom bomb in the form of two hemispheres of enriched uranium, using a miniature gun to shoot one of these hemispheres, as a projectile, into the other hemisphere, which served as a target. As long as these hemispheres were apart, the bomb was inert; when they came together the enriched uranium sphere would exceed the critical mass and the bomb would explode. All of this seemed to them exceedingly simple and would require a staff of the order of 50 scientists and maybe up to 50 additional assistants.

Oppenheimer proposed that the project be set up near his summer home in a remote valley near Las Vegas, New Mexico. However, this would require a considerable amount of new construction and thereby delay the research. Dr. Conant then suggested taking over the Los Alamos Ranch School, a very elite preparatory school for boys where he had considered sending his son (tuition $3500). This already had facilities for 50 boys and 30 teachers. There were also two small working ranches in the area. Together they seemed ideal for the project. Los Alamos is high up on a mesa with the Rio Grande River to the east and Jemez Mountains to the west. The difficulty was that none of the California physicists had had any experience with ordnance. Dr. Vannevar Bush brought in Commander (subsequently Admiral) W. S. (Deak) Parsons to take charge of the ordnance work. Parsons promptly announced that he would need laboratory and range facilities for 200 ordnance experts. This started the large-scale expansion of the Los Alamos personnel and facilities. The next step came when the physicists realized that they would need to measure cross-sections for a variety of nuclear reactions. This led to an invitation to Bob Wilson to bring his group of experimental physicists together with their high pressure Van de Graaff generator from Princeton. The next major change at Los Alamos occurred when a University of Washington physicist, Seth Neddermeyer convinced John Von Neumann, Hans Bethe, and Robert Oppenheimer that the best way to construct a plutonium atomic bomb was to use high explosives to implode a metal shell. This required the efforts of George Kistiakowsky and the group of high explosive experts which had been working with him on conventional bombs at the Bruceton Laboratory of the National Defense Research Committee. Thus, the size and scope of the Los Alamos project rapidly mushroomed. Hans Bethe headed the theoretical research; Cyril Smith brought in a large number of metallurgists and chemists; Sir James Chadwick brought with him a very distinguished group of British scientists; etc.

The idea was that as each new problem developed, experts were recruited to solve that problem. For example, I was brought in to prescribe the characteristics of the gun and gunpowder to be used in the enriched uranium atom bomb. This was easy for me since Charles Curtiss, Dick Kershner, and I had just developed a very general system of internal ballistics for guns and rockets which was compatible with the laws of physics — all previous systems were semi-empirical. The new feature in our interior ballistics was the inclusion of heat transfer from the powder gas to the bore of the gun; we used Von Karman's formula for convective heat transfer in pipes, as given in McAdams' well-known elementary chemical engineering text. For a large Navy cannon, the heat transfer was unimportant, but for a high-powered military rifle, approximately 15 times as much of the powder gas energy is expended in heating the bore as in pushing the bullet. Dr. Richard Tolman was General Groves' scientific advisor and familiar with my work. He discussed with me the general requirements for the Los Alamos gun without telling me what it would be used for. I recognized the similarity of this gun with a super-secret device which the British were developing to penetrate the 15 feet of concrete which protected German submarine pens; thus, I did not make the connection between Tolman and the atomic energy project. I made preliminary calculations and sketches, after which Dr. Tolman introduced me to General Groves; the next day I flew to Los Alamos. In three weeks I completed the final specifications for the internal ballistics of the gun and the charge of gun powder. Because of security, it would have been unwise to let me go back to my previous work in Washington. Thus, I was assigned to an entirely different sort of job; I was made a group leader in Hans Bethe's Theoretical Division and my group was given the problem of determining all of the effects of an atomic bomb which would take place after the nuclear fissions had occurred. Lest there be any confusion, in the Los Alamos hierarchy a division leader such as Bethe was a big shot, whereas group leaders such as myself were little shots. After all, just before the war I was an assistant professor at the University of Wisconsin and quite young.[2] Because Oppenheimer insisted on the laboratory being democratic, group leaders were included in the frequent Coordination Council meetings where the technical decisions were made. And, although each group had its primary responsibility, we were free to call upon anyone in the laboratory for help. Many of the weapon effect problems were completely unfamiliar to me and I was most grateful for the unselfish help which we received from many people. Wouldn't it be wonderful if we could all work together in this manner for peaceful instead of military objectives?

My situation at the end of the first three weeks represented a typical personnel problem at Los Alamos: After the expert had solved his problem, what did you do with him? Generally the expert was transferred to some other problem area where he might, or might not, be knowledgeable. However, there remained the question as to whether he should be placed under or over a younger and less distinguished man who really understood what he was doing. In either case, personnel problems were bound to develop. By the time that the atom bomb was tested at Alamogordo, Los Alamos was three deep in experts and tensions were so high that if the atom bomb had not functioned, there would have been a much greater explosion between the scientists.

Los Alamos was a strange mix of military and civilian personnel. Most of the senior scientists were civilians. General Groves preferred that the scientists be civilians because this avoided all sorts of red tape, and, of course, scientific ideas have no regard for either age or rank. However, civilians required higher salaries and rated excellent apartments or dormitory accommodations, whereas GIs were housed in inexpensive barracks. In the spring of 1943, just before I went to Los Alamos, I received a commission in the Army to serve in the front lines and report on the functioning of our conventional guns and rockets. General Groves made me refuse this commission and told me not to give any explanation. You should have heard the dressing-down that the officer in the Pentagon gave me when I told him that I had decided not to serve in the Army.

At Los Alamos very few chemists, physicists, or engineers were able to pursue their specialties very long. We were all seeking solutions to difficult problems and our research was not limited to areas where we had had previous experience or training. For example, let me sketch the large number of seemingly unrelated problems that John Magee and I had to study in order to predict the radioactive fallout to be expected from an atomic bomb explosion:

(1) First, we had to study the formation of the ball of fire. This involved aerodynamics, including the radiative transport of energy, and for this we studied Chandrasekhar's astrophysics treatise on the dynamics of nebulae and chemical engineering books such as McAdams' *Heat Transport.*

(2) The second problem was the rise of the fire ball and the generation of winds along the ground and surrounding the fire ball. Sir Geoffrey I. Taylor, the world's greatest authority on turbulent and convective aerodynamics, helped us with this problem. He scribbled the key equations onto the back of envelopes, then John Magee and I carried out the necessary computations.

(3) Next we had to figure out the fate of the radioactive materials inside the fire ball. What size distribution of the particles is to be expected? There were two possibilities depending upon whether the fire ball was sufficiently high above the terrain that debris from the ground did not mix with the bomb materials. If the ball of fire were high above the ground, small particles of the radioactive bomb material condensed on the interface between the fire ball and the cold air currents surrounding it. For this, we studied Volmer's book on *Homogeneous Nucleation*. However, the commercial formation of carbon black for use in rubber tires provided us with a more relevant treatment and a wealth of useful data. Magee had worked in the Goodrich Rubber Company laboratory at the beginning of World War II, and, therefore, knew how to get access to the necessary documents. On the basis of the Volmer theory and the carbon black, we convinced ourselves that homogeneous nucleation could only produce very small radioactive particles. However, to clinch the matter, John and I made experiments in which we vaporized metals and chilled the vapor by mixing with cold air. Sure enough, we got a blue smoke of metal particles less than a micron in diameter, in agreement with our theory.

(4) If the bomb were detonated within a half-mile of the ground, the air currents sweeping along the ground pick up dirt, rocks, and assorted debris, some of which pass through the fire ball and get plated with radioactive materials. To learn how wind picks up debris from the ground, we studied Department of Agriculture Soil Conservation reports and Bagnard's treatise, *The Physics of the Blown Sand* (a study made in the Sahara Desert). The thing that is amazing is that the size distribution of these particles depends very, very little on the kind of terrain that this crud had been picked up from.

(5) Having determined the initial size distribution of the radioactive particles, it remained to study how these particles would spread out and fall with the passage of time. This led us to a book entitled *Micromeritics* which is a study of air pollution from industrial smokes and particulates emitted from smokestacks. Of course we also studied the spreading out of airplane trails, etc. The biggest difficulty is in estimating the effect of gusts and turbulence in the upper air. In order to learn what happens to the particles as they come close to the ground, we studied Chemical Warfare reports on how poison gases distribute themselves with respect to hills and valleys. Most of the radioactive fallout occurs in the valleys, leaving the hilltops relatively uncontaminated. At the Alamogordo bomb test this was important because most of the people in the area lived on hilltops, so they were not in danger. However, their cattle spent most of their time grazing in the valleys and quite a few cows

suffered from skin tumors. Since cows don't bathe and can't wash off this radioactive dust, people have an additional advantage.

(6) There were many more problems. For example, we consulted with the top military meteorologists with respect to air circulation in the upper atmosphere for distances of thousands of miles. Our biggest concern was the washing out of the radioactive particles by rain, since that was not predictable. After the Alamogordo bomb test there was rain in Illinois which brought down a small amount of fallout which was sufficiently radioactive to make the wheat shafts unsuited for use in Eastman Kodak photographic films. I wonder how Eastman Kodak produces photographic films when literally all of the surface of the earth, at least of the northern hemisphere, has a certain amount of radioactive contamination.

It would be difficult to carry out such interdisciplinary research in a university where we are all divided up into departments where most poeple are working on closely related problems. In industrial or government laboratories, interdisciplinary problems are solved by task forces composed of people having different skills and backgrounds. Frankly, I am very much concerned that the training which we give our students is so highly specialized that they are not prepared to tackle problems which are not closely connected with their theses. It is important that our students develop sufficient breadth that they can explain their ideas to people with different backgrounds. This is essential if they are to become useful members of an interdisciplinary task force.

In spite of all this work, very few people believed us when we predicted radiation fallout from the atom bomb. On the other hand, they did not dare to ignore this possibility. Thus, at the Alamogordo (or 'Trinity') bomb test, Dr. Stafford Warren (a radiologist who was subsequently Dean of Medicine at UCLA) was put in charge of the fallout operations and John Magee and I were his chief helpers. However, we had such low priority that the best transportation we could get was an old automobile which we borrowed from Jim Tuck. On the evening before the test, just after sundown, John and I left Los Alamos and drove to the La Fonda Hotel in Santa Fe where we contacted some Manhattan Project detectives who gave us the fancy three-scale Geiger counters which we were to use the next day. (Geiger counters were regarded as super secret since their presence would link Los Alamos with nuclear radiation. Thus, their shipment and delivery was carefully guarded.) Since it was pitch dark, we did not get a chance to try them out. Then we drove to the Hilton Hotel in Albuquerque where Robert Oppenheimer was meeting with a large group of generals, Nobel laureates, and other VIP's. Robert was very nervous. He told John and myself about some experimental results which Ed Creutz had obtained earlier in the day which indicated that the atom bomb would be a dud. Because President Truman needed to know the outcome of Trinity to use in his Potsdam negotiations with Stalin and Churchill the following day, there was no possibility of postponing the test. After a hamburger and a cup of coffee, we drove through the rain the remaining 150 miles to the Journey of Death Valley where the atom bomb was to be detonated. We arrived there around two o'clock in the morning and joined up with the convoy of soldiers whom we were going to lead the next morning. Our post was 10 miles from ground zero. Staff Warren was with the high command at another post 6 miles from ground zero. His job was to deploy a number of security agents to monitor the habitations downwind. Unfortunately, he misinterpreted the wind direction and the radiation monitors were deployed *upwind*. This became apparent in talking to him through my unreliable walkie-talkie. By the time that this got straightened out, it was too late to move the monitors to places where they would be useful. At this point, I took a catnap in the automobile while the cold rain continued to fall. Jack Hubbard, the meteorologist, had predicted good weather. When the rain started, he locked himself up in his room and prayed. His prayers must have been potent because the rain stopped only minutes before blast-off.

When I woke up it was time to get ready for the explosion. There were 300 of us assembled at our post. These included soldiers, scientists, visiting dignitaries, etc. We were all cold and tired and very, very nervous. Most of us paced up and down. I was fascinated with the preparations which Fermi was making to measure the energy yield of the atom bomb by the horizontal distance which a paper strip traveled when released from a height of four feet. Fermi had made the calculations on the basis of our shock-wave wind velocity versus distance charts (which scaled as the one-third power of the energy yield). We all had been given special very, very dark glasses to watch the explosion. However, John and I had decided to turn our heads away from the explosion and not to try to look at it for the first one or two seconds. The explosion took place just before dawn with the sky still dark. All of a sudden, the night turned into day, and it was tremendously bright, the chill turned into warmth; the fire ball gradually turned from white to yellow to red as it grew in size and climbed in the sky; after about five seconds the darkness returned but with the sky and the air filled with a purple glow, just as though we were surrounded by an aurora borealis. For a matter of minutes we could follow the clouds containing radioactivity, which continued to glow with stria of this ethereal purple. There weren't any agnostics watching this stupendous demonstration. Each, in his own way, knew that God had spoken. We stood there in awe as the blast wave picked up chunks of dirt from the desert soil and soon passed us by. We all felt relieved that the bomb had exploded, but we would have been much happier if we had been able to prove that no atom bombs could ever be made. If atom bombs were feasible, then we were glad that it was we, and not our enemy, who had succeeded.

Fermi's paper strip showed that, in agreement with the expectation of the Theoretical Division, the energy yield of the atom bomb was equivalent to 20,000 tons of TNT.[3] For a period of a few hours, this was the best measurement of the energy yield. It is noteworthy that Professor Rabi, a frequent visitor to Los Alamos, won the pool on what the energy yield would be – he bet on the calculations of the Theoretical Division! None of us dared to make such a guess because we knew all of the guesstimates that went into the calculations and the tremendous precision which was required in the fabrication of the bomb.

Right after the blast wave passed by, John Magee and I started the engine of our old automobile and led the convoy of soldiers into the desert. Our mission was to follow the radioactive clouds, to monitor the fallout from these clouds, and if the radioactivity was sufficiently high we were prepared to evacuate the desert dwellers. About 25 miles from ground zero, we came upon a mule who must have looked directly at the explosion – his jaws were wide open, his tongue hanging down, and he was completely paralyzed. When we passed the same spot in the afternoon, the mule was gone so he must have recovered. Then we came to a small store at the crossing of two dirt roads. John and I rang the door bell and an old man came out. He looked quizzically at us (John and I were wearing white coveralls with gas masks hanging from our necks). Then he laughed and said, "You boys must have been up to something this morning. The sun came up in the west and went on down again." There was some fallout there, but the level of radioactivity was not dangerous. The soldiers bought almost everything in the store and left his shelves bare. (One would have thought that the soldiers could buy everything they had wanted at the Army PX, and therefore, a country store would not have had such an appeal.) Our next stop was a visit to an Army searchlight post. The soldiers there had bought some huge T-bone steaks to eat after the atom bomb explosion. When we arrived they were just roasting the meat and it smelled delicious. However, at the same time, the fallout arrived – small flaky dust particles gently settling on the ground. The radiation level was quite high. So we sent the searchlight crew back to their base camp and told them to bury their steaks. Then we drove into a valley where the radiation level was

20 Roentgens per hour which is definitely dangerous for anyone remaining there for more than a short time (1–2 hours). And finally we headed back. On the way we had to change a radioactive flat tire. When we reached our base camp our whole automobile was so radioactive that the Geiger counters read 4 Roentgens per hour in the driver's seat. (Later, the soldiers scrubbed the car and cleaned it up in every way possible. Nevertheless, four days after the explosion, when the car was driven up to Los Alamos, it still was sufficiently radioactive to throw the Geiger counters in nearby laboratories off scale.) John and I scrubbed ourselves with lots of soap and repeated rinsing and changed into clean clothes. General Groves thanked us for our efforts and suggested to a Major General that he drive us to Albuquerque, where we would spend the night before returning to Los Alamos. We started out by sitting next to the Major General but it was clear that he was afraid that he would get radiation sickness from being exposed to us. First, he asked us to sit in the front seat with the GI driver. Then he hailed a jeep and ordered the soldier driving it to take us the remaining 130 miles to a hotel in Albuquerque. (I have seen this Major General a number of times since then and he greets me warmly and reminds me of the exciting times we had together at Trinity!) Gosh, we were glad to go right to bed when we arrived.

In the span of thirty years I have forgotten a great many details about the early days of Los Alamos. However, I will never forget some of the great men with whom I had the opportunity of working. Let me tell you about a few of them:

J. ROBERT OPPENHEIMER

Every Sunday he would ride his beautiful chestnut horse from the cavalry stable at the east side of town to the mountain trails on the west side of town greeting each of the people he passed with a wave of his pork-pie hat and a friendly remark. He knew everyone who lived in Los Alamos, from the top scientists to the children of the Spanish-American janitors — they were all Oppenheimer's family. His office door was always open and each of us could walk in, sit on his desk, and tell him how we thought that something could be improved. Oppy would listen attentively, argue with us, and sometimes dress us down with clever cutting sarcasm. At all times he knew exactly what each one of us was working on, sometimes having a better grasp of what we were doing than we did ourselves. Needless to say, we all adored and worshipped him. During the first year at Los Alamos, he was a wonderful director. However, after the personnel at Los Alamos expanded to more than 300 scientists, it became necessary for him to delegate authority to an associate director — and this didn't work out very well. It takes a different kind of a person to administer a large organization than a small one.

Oppy was a frail, sensitive, lonely man whose life was tragic. When he was thirteen years old, he had tuberculosis and his parents took him to New Mexico where they bought a ranch in a mountain valley far from any habitation. There, Oppy whiled away the time while recuperating by studying Sanskrit. He was absolutely brilliant. However, in spite of Oppy's eminence as a physicist he published very few articles; indeed, he set such high standards for his own research that he could seldom satisfy his own standards. This made him unhappy. Oppy had a wonderfully clear understanding of the basic principles of physics and had a genius for finding other people's mistakes. Thus, physicists from all over the world came to him for advice and help.

Oppy felt that even the lowliest janitor might contribute a key idea to the Los Alamos project. Therefore, he insisted that (as much as possible) within the confines of the community everyone should have access to all of the information. This was one of the reasons why Los Alamos had to be located in such an isolated spot and why our mail, travel, etc. had to be subject to severe security regulations. The security was successful as far as the Germans were concerned — Hitler never found out about Los Alamos. However, Klaus Fuchs did succeed in passing most of the Los Alamos secrets to the Russians. Somehow or other the townspeople in Santa Fe got the idea that we were working on some kind of submarine. Why, I don't know. Some scientist must have told that as a joke. But I suppose that as far as rumors are concerned, they don't have to be logical to get wide distribution.

All through the war the security guards left a little hole in the fence so that the Indians could climb through and come to the moving pictures (admission 12¢), and also do their shopping in the PX. I think Oppy was responsible for that hole.

NIELS BOHR

A very distinguished looking old man and his son appeared at Los Alamos. Judging from the number of security guards which were accompanying them and the deference that Mr. Achers, the manager of Fuller Lodge (which offered guest accommodations and board to official visitors) showed, this man must be a big shot. He said that his name was Nicholas Baker and he was very much afraid lest people recognize him as Niels Bohr — I think his wife was still in Denmark. Bohr had just had a very frightening experience. He had been whisked out of Denmark and taken by boat to Sweden. Then he was flown in a small military plane to England. He was placed in the bomb bay of this plane and provided with earphones so that the pilot could talk to him. Somehow or other, he did not hear the pilot tell him to put on an oxygen mask when the plane climbed above the German antiaircraft searchlights, etc. Thus, when the plane arrived in England, Bohr was unconscious. Since Bohr's forehead was so high, I have often wondered whether the earphones were long enough to reach his ears. Bohr and his son, Aage (the 1975 Nobel laureate in physics), were almost inseparable. Each day they would take a long walk during which they would discuss some very difficult physics problem. Niel's ideas were always extremely simple and extremely basic; he never got confused because he always bore in mind the basic premises and the approximations which were being assumed. He reminded me of the quotation, "God and Nature are simple, it is we who are complicated!" I always had the feeling that Bohr was very close to God.

Bohr was very lonesome and occasionally he would come over to our apartment for some of my mother's homecooked food. One time we took him on a picnic up in the mountains. He had a wonderful time and proclaimed that that was the first picnic that he had ever had in his life (this I never understood). I will never forget his happy smile while he was eating a big slice of watermelon. Actually this is the only occasion where I ever saw Niels Bohr thoroughly relaxed. Even before World War II, Bohr always had a very sad expression and looked as though he carried all of the cares of the world on his shoulders.

Niels Bohr is famous for playing a major role in the development of quantum mechanics. In addition to his own research contributions, a whole generation of physicists came to his Institute in Copenhagen to seek his advice. Thus, he taught a whole generation of physicists how to change their way of thinking so as to cope with the uncertainty principle and the dual particle-wave nature of matter.

It is surprising that Bohr had great difficulty in comprehending the effects of the atomic bomb. He understood the nuclear reaction involved and he expected that all of the energy of the atom bomb would come out in the form of neutrons and radiation. He found the formation of the shock (or blast) wave puzzling. Most of the thermal radiation emitted is in the form of very long wavelength X-rays or very short wave ultra-ultraviolet light. Cold air is almost completely opaque to radiation in this wavelength region. Thus, this thermal radiation is trapped in the fire ball. The trapping process leads to the building up of a large pressure difference between the fire ball and the surrounding air. After which the pressure wave separates from the ball of fire and the blast wave progresses as in a conventional explosion.

John Magee and I worked out the theory of the blast wave under the direction of Hans Bethe and with the help of John Von Neumann, Sir Geoffrey Taylor, Sir Bill Penney, Klaus Fuchs, and a large number of less famous men and women (the women were either WACs or wives, for example, Mrs. Teller, Mrs. Manley, and Mrs. Brode. They carried out computations on desk calculating machines and did some work using the primitive IBM computing units: collating, adding, multiplying, etc. These units were not connected together to form a digital computer in the sense that we now know it.). The large number of people who helped in this problem is typical of the massive effort which was involved in the Los Alamos project. In a very literal sense, it was a team effort and it is difficult to assign credit to any one individual for any particular development.

HANS BETHE

I learned a great deal from working under the direction of Hans Bethe. Like all great physicists, he had a marvelous intuition based upon wide experience in tackling all sorts of different problems. As Einstein said, "An expert is a person who has made every mistake once." In order to treat the aerodynamics of very hot gases, Bethe devised a new type of perturbation scheme (where $\gamma - 1$ is the perturbation parameter, γ being the specific heat ratio). This was very useful under some conditions. Bethe was smart enough also to figure out where other methods would be more appropriate. What impressed me most was that Bethe not only was ingenious in approximating the solution to complicated equations, but he also determined whether the calculations gave values too large or too small and what was their probable accuracy.

Unlike Oppenheimer or Bohr, Bethe was a jolly fellow. My office was down the hall, and I could always tell when Bethe and Feynman were working together because there would be frequent bursts of guffaws and belly laughs. When there was a comic moving picture in the base theater, you could always tell by the laughter whether Bethe was in the audience.

Bethe's hobby was mountain climbing (with and without skis). Very often on a Sunday, Bethe would climb up to the top of Lake Peak (12 500 feet) with Enrico Fermi or some of his other friends and sit there in the sunshine discussing physics problems. This is how many discoveries were made.

It gives me a comfortable feeling to know that Hans Bethe is one of the top Americans in the SALT talks – I trust his good judgment.

ENRICO FERMI

Fermi felt very strongly that every physicist should be able to make back-of-the-envelope calculations of anything and obtain an answer which is correct to within an order-of-magnitude. For example, at tea one afternoon, he asked his associates to estimate the number of railroad locomotives in the United States. A typical solution would involve estimating the number of miles, on the average, that you would have to drive before you crossed a railroad track; from this, you obtain the total number of miles of railroad in the United States; then you estimate the number of miles of track per locomotive. By stressing the orders-of-magnitude of things, Fermi wanted his boys to understand what is important and what is negligible. He felt that we learn all of the complicating factors in the problems that we study, but we seldom take the time to evaluate their relative significance.

Fermi, like Bethe, was an ardent mountain climber. At Los Alamos he was usually accompanied by his bodyguard, a very nice Italian-American Chicago lawyer. The poor bodyguard would generally arrive in Los Alamos with only his street clothes. When Fermi climbed Lake Peak or one of the peaks behind Los Alamos, the guard would trudge along. I can remember this guy coming back to Fuller Lodge, thoroughly bushed, mumbling, " . . . and just to think he could have ridden a horse!"

ADMIRAL WILLIAM S. PARSONS

My candidate for the UNSUNG HERO of Los Alamos is Admiral Deak Parsons. He was the Director of Ordnance Division and one of my bosses. Governmental and military high commands in Washington respected the judgment of Admiral Parsons more than anyone else in Los Alamos. Before working on the atom bomb, Parsons introduced radar to the South Pacific — the Japanese had become accustomed to sailing their ships through a narrow strait on foggy nights. On this particular foggy night, Parson's radar spotted seven warships and they were sunk. At Annapolis as a midshipman, he suggested a number of ways that the standard procedures for firing cannons could be improved. All of his life he fought the silly regulations and the conservatism of the Navy. Before risking a confrontation, Deak would patiently prepare an ironclad case; in this way, he always won! It was great fun working for Deak. He backed you to the hilt and took full responsibility for your boners. After graduating from the Naval Post Graduate School, Parsons spent many years at the Naval Ordnance Proving Grounds at Dahlgren where he became a protégé of Dr. L. T. E. Thomson, a physicist who had helped Goddard develop the first American rockets. Thomson persuaded the three great physicists, Eugene Wigner, Fred Seitz, and Ralph Sawyer, to spend their summers at Dahlgren and incidentally enjoy the sailing on the Potomac River and the wonderful parties which made Dahlgren famous. These physicists taught Parsons and it opened up a wonderful new world to him. Every night, even while he was at Los Alamos, he would read a new physics treatise in much the same way that most people read a detective story — he found it exciting and he learned rapidly. At Los Alamos he was never awed by the great physicists. Whenever one of them would state a conclusion, Parsons made him prove it. Indeed, I can remember an occasion when Parsons made Niels Bohr admit that his basic assumption had not been carefully considered, and another time when Parsons made Bethe back down. After World War II, Parsons was rapidly promoted to Assistant Chief of Naval Operations. His last assignment was to head up a committee to revise the military academies — he was very much looking forward to replacing the military by civilian professors and raising the standards to become comparable with those of the best universities. Unfortunately, Admiral Parsons died of a heart attack at the age of 43. I can still remember Parsons' comment about Los Alamos, "Isn't it wonderful to see 5000 people intent on a single purpose — skiiing!"

I have many fond memories of Los Alamos. We worked hard, we played hard, we made many friends, we climbed the surrounding mountains, we got acquainted with the Indians, and we learned a great deal. Los Alamos is an example of how a scientific-technological miracle can be performed!

DISCUSSION

Question: You said that it took a genius to figure out the ramifications of solving a problem. What do you think is going to happen in the future if you form more task forces? Who is the genius who's going to figure out the consequence of the resulting discoveries?

Hirschfelder: Frankly, I am quite amazed at the success that we had in predicting the effects of the atomic bomb. I would say that we didn't miss a single effect, except perhaps the base surge at the Bikini bomb test underwater explosion. We came within 10% on each of the different effects. I went through in considerable detail the kind of thoroughness that one had to have in order to make these predictions and to figure out what these effects were going to be. We have the theoretical ability; we have the computing machines; all we are lacking are the research funds and the commitment of the government to support the long-range research which is required for such work.

Question: Do you feel that the research results of a major project should be held back from either public announcement or public use until its effects can be properly estimated?

Hirschfelder: Let me speak about what happened on the atom bomb. Immediately after Nagasaki, John Von Neumann got a group of us together with the idea that it was essential that we tell the complete scientific story as far as we could without telling the production secrets of how you build an atom bomb. And the result of our meeting was that I was selected to become senior editor for an unclassified treatise on the effects of atomic weapons. I set out immediately to get together all of the experts on the effects of the atomic weapon. There were 237 experts that I got to help with this writing, so it wasn't just an individual effort. These 237 experts discussed each of the different kinds of effects. Then we had the job of going through the usual Atomic Energy Commission security channels. Every part of this had to be approved for release by the AEC Senior Technical Reviewers. This is what finally came out. However, I had a running battle with the Public Relations Division of the Atomic Energy Commission. At one time I received a telegram from them that was booby-trapped in such a way that if I had answered the questions asked of me in this telegram, they would have clapped me in jail. There was a tremendous amount of opposition to releasing our book until the Russians exploded their atom bomb. When the Russians exploded their atom bomb, the public relations men who had been blocking our efforts went to a number of my chapter editors, got from them their first drafts of manuscripts, and published these first drafts without any security clearance whatsoever. So my answer is that it is very important that, as much as possible, information be made available and complete to the public. My feeling has always been that as far as the principles and discoveries in physics are concerned, these can be concealed maybe for a month, maybe for a year, but not for any long time. When the time is ripe to make a discovery in physics, there are half-a-dozen people who are going to make that discovery. However, as far as ordnance is concerned, if you put together a fiendish contraption, the only people that need to know are those that want to make a fiendish contraption, and I don't see any reason to make things easier for them.

Question: How do you feel about the possibility of a person hijacking or getting hold of some radioactive material and making a bomb and possibly using it for blackmail or fiendish purposes?

Hirschfelder: Such hijacking might be good for blackmail. However I doubt that they could make an atom bomb. I can assure you that they would make an awful mess. Plutonium is nasty stuff. Its radioactivity is in the form of alpha rays. Gamma rays and beta rays can more easily be detected with a

Geiger counter. Alpha particles are much harder to detect with the equipment usually available. There is some evidence that if you inhale plutonium, even one little particle of the right size, it can get stuck in the lung, fester, and produce cancer. So it's nasty stuff. A homemade bomb might dirty up an area of maybe 100 yards. The poor guys that go in there to clean up the mess might have some difficult problems unless they knew what they were up against.

Question: Do you think the safeguards for this type of material are adequate?
Hirschfelder: There is no excuse why our expensive radioactive materials should not be properly safeguarded. I read an article in the newspaper the other day that stated that a large amount of nuclear materials was unaccounted for. While most of that undoubtedly is in the pipes of the gaseous diffusion plants and the interior of other equipment, the AEC (now NRC) cannot tell if significant quantities have been stolen.

During the war there were quite a few items stolen from the Los Alamos laboratory. The most brazen example was a mechanician who drove down to Santa Fe and arranged to sell his turret lathe to a man he met at the La Fonda Hotel bar. The mechanician returned to Los Alamos, unscrewed his turret lathe from the floor, used a portable crane to move it to the loading dock and placed it on a 6 X 6 Army truck which he had borrowed, covered it with a tarpaulin, drove it past two guard gates unchallenged, drove it down to Santa Fe. Finally, when he delivered the lathe to the man at the La Fonda bar, the mechanician was arrested — the man at the bar was an undercover Army detective!

Question: What was your and your associates' reactions to the announcement of the Soviet Union's atomic bomb?
Hirschfelder: Of course, it was expected that sooner or later the Russians would have an atom bomb, but it actually occurred about two years sooner than we had expected. The role of Klaus Fuchs, the arch spy, who was located at Los Alamos, probably shortened the Russian atom bomb development by two years. Klaus Fuchs was an extremely able physicist. He ranked in stature with people like Hans Bethe and Robert Oppenheimer, or Edward Teller. On one of the basic patents for the hydrogen bomb, you'll find the name of Klaus Fuchs. He was a very clever guy. He set out to discover what was going on in the Manhattan Project. He very systematically went to each of the laboratories in the Manhattan Project, got himself placed in a key position, and transmitted this information. For example, at Los Alamos, he ended up

as the editor of the 25-volume secret Los Alamos Encyclopedia which summarizes all of the research which was carried out at Los Alamos. He hated chemists. He always stuck his nose up at them. He was a very shy fellow but a good babysitter: he used to babysit for Edward Teller's little boy. He would never want to be off in a corner with you alone. He liked to be in a group where there was small talk. Of course, we never suspected that he was a spy.

Question: Could you give us a picture of how the group reacted when Oppenheimer was accused of disloyalty in the early 1950's and lost his security clearance?

Hirschfelder: This is a very complicated question. Very complicated because Oppenheimer himself was a very complicated guy, and almost anything that is said about Oppenheimer is true some times and not true other times. Of course, nobody thought of Oppenheimer as a man who would intentionally violate security or anything like that. However, Oppenheimer as director of the Los Alamos Laboratory did not intervene when young people in the laboratory who had violated security were given harsh kinds of punishment. For example, every once in a while there would be some young loud-mouth who would go down to the La Fonda Bar and shoot his mouth off; there would be detectives from the Manhattan Project with notebooks taking this all down; and they would build up a book on the indiscretions of a particular guy. The young loud-mouth wasn't warned, but once they had the book on him, enough so that they could salt him away, he just disappeared and we always believed (I don't know whether it's true or not) that he was sent off to some isolated island where he would be incommunicado for a long length of time. I can cite examples where Oppenheimer did not turn a finger to help a number of these young guys who were being accused. This just wasn't his business. So, when Oppy himself committed the indiscretion of lying to security agents, it seemed reasonable for him to have to obey the same laws as the people that he directed. But as far as the timing was concerned, it was terrible. I mean if Oppenheimer had been regarded as a security risk, he should have lost his clearance immediately, not four or five years later. It so happened that just before he lost his clearance, the Atomic Energy Commission had changed their security rules so as to permit taking under consideration the security of your family. Of course, Oppy's brother Frank and his wife Kitty could be regarded as security risks. However, Oppy had enemies. We scientists regarded Oppenheimer as exceedingly clever when he cut us down with his sarcasm, but the military guys couldn't stand losing face and so there were a number of them, and one in particular, who were out to get him. The whole thing was just tragic and unfortunate. I admired Oppy tremendously and hated to see him suffer.

Question: Was there any consideration on the part of the scientists about whether the bomb should be used in World War II?
Hirschfelder: Amongst the scientists at Chicago, there was a great deal of consideration along those lines. At Los Alamos we never questioned that it should be used in a military engagement. For two reasons: First of all, if the atom bomb is exploded high above the ground (as it was both at Hiroshima and Nagasaki) very few of the casualties are produced by radiation – most of the casualties are produced by falling rubble, blast, and fire as in conventional air attacks. Secondly, the number of casualties and the extent of the damage to structures at Hiroshima and Nagasaki were comparable to the conventional air attacks on Tokyo, Hamburg, or Dresden. Although the energy yield of the atom bombs were 20,000 tons of TNT, in point of fact the damage and casualties which they produced was equivalent to 1000 two-ton block busters. This is because the area in which the casualties and damage occurs varies as the two-thirds power of the energy released.

Question: What was it like to have your mail censored?
Hirschfelder: At first our mail arrived opened up and obviously censored, but nobody took responsibility. Then the Army Engineers officially censored the mail. The civilians sometimes played jokes on the censors – for example, one of Julian Mack's group received a letter: "Dear John: I didn't know what to buy for your birthday so I enclose a $5 bill." No $5 bill was enclosed. The censor went to his superior and asked what to do about it; the superior told him to put a $5 bill into the letter so as to avoid any difficulties with the civilians. Soon thereafter, the censors were faced with many similar letters!

Our mailing addresses were: P. O. Box 1663, Santa Fe, New Mexico for personal correspondence; Los Angeles for technical publications; Washington for some technical correspondence; etc. Because all of our friends used the Santa Fe P. O. Box address for personal correspondence, this complicated system fooled nobody.

Question: Were the biological effects of the atom bomb improperly assessed?
Hirschfelder: No, the biological effects were quite accurately assessed.

Question: Then you knew that it would cause many deaths because of radiation poisoning as well as because of the blast?

Hirschfelder: Actually, in Hiroshima and Nagasaki there was no residual radioactivity left on the ground. The height of the detonation was so high that the fire ball did not come in contact with the ground. The result was that the radiation exposure of people was mainly gamma rays. If people were in a region where they got a dose of neutrons, they were killed by the blast and by fires and by all sorts of other things as well as by radiation. At Hiroshima and Nagasaki, I do not believe that there were more than one or two thousand people who were left with radiation burns (but not killed). It is a surprisingly small number.

Question: Were the long-term effects of radiation assessed?
Hirschfelder: These thousand or two thousand people who had suffered the radiation exposure between the range of 100 Roentgens to 500 Roentgens *are* now having problems. As I understand, their grandchildren may have problems. Their children are probably less likely to have problems than their grandchildren. These things are tragic, of course they're tragic, and war is hell. There isn't any humane way of fighting a war; war is intentionally killing people. One of the advantages of atomic warfare over that of the previous warfare is that it is much more democratic; civilians and everybody have a chance of getting killed. There aren't any people that can be protected. As far as future warfare is concerned, a hydrogen bomb can be a thousand times the strength of our World War II atomic bombs. Something like a hundred hydrogen bombs would certainly be sufficient to make any country come to its heel. We have reached the point where we can annihilate the world and make it uninhabitable by civilized people. This is not a sensible way of solving any disputes. We just cannot afford to have a nuclear war. There is no protection from nuclear warfare.

NOTES

[1] In order to make the plutonium piles at Hanford so much carbon black was required that for a period there was a shortage of rubber tires in the United States.
[2] All through the war my salary was based upon my assistant professor status.
[3] I had coined the unit of one ton of TNT equal to 10^9 calories rather facetiously. Unlike most explosives, the energy yield of TNT depends upon the conditions under which it is exploded. Thus, I could define a standard yield of TNT in nice round numbers and be sure that there would be some condition of firing such that the real TNT would have an energy yield which agreed with my standard!

LAURA FERMI

THE FERMIS' PATH TO LOS ALAMOS

One problem facing the amateur historian is to find a good starting point. When I was writing the story of atomic energy for young people, I hired a ten-year-old consultant. One of his great contributions was to prevent me from leaving out of the book the part on Democritus and his ancient colleagues. An exhaustive history of Los Alamos ought to start with Democritus. But since I am not qualified to relate what I have not seen, the story must begin in Rome, Italy, where I was born and raised. If I look at the personal events that eventually took me to Los Alamos I find that I can choose two starting points. One is scientific and the other is political. The scientific date is the middle of January, 1934, when Fermi came back from his skiing vacation in the Alps and found in his mail an announcement by the Joliot-Curies in Paris that they had achieved artificial radioactivity by bombarding a few elements with alpha particles. Fermi was then mainly a theoretical physicist, but theoretical physics had just caused him disappointment and irritation; the journal *Nature* had refused to publish his paper on the theory of beta decay, which is still considered as one of his main works, on the grounds that it seemed crazy. So he welcomed the chance to shift to experimental physics and take up artificial radioactivity. He thought that neutrons might be more effective than alpha particles to induce radioactivity, but to be sure he had to try.

The political and probably more important starting point is earlier: October 28, 1922, the day of the Fascists' march on Rome, which brought Mussolini to power. His policies, the encouragement and the example he gave Hitler, and his later subservience to the Germans are the causes that led the Fermis to emigrate and join that huge intellectual migratory wave that came from Europe in the thirties. I mean the men and women fully educated in Europe, who found their way to America after having lost their positions or given them up because of the racism and political intransigency of the European dictators. Arriving, as they did, with their previously acquired cultural baggage and well-formed mental attitudes, they eventually changed the American cultural scene.

Men like Albert Einstein and Thomas Mann, who were preceded by their reputation, gave prestige to American science and letters, respectively, by their

L. Badash, J. O. Hirschfelder and H. P. Broida (eds.), Reminiscences of Los Alamos
1943–1945, 89–103.

mere presence. Others affected American life more directly. The Austrian
psychoanalysts spread the Freudian word and deeply influenced our way of
thinking. Musicians like Igor Stravinsky. Arturo Toscanini, George Szell, and
Rudolf Serkin molded the American musical taste. There were artists, scholars
and scientists of all kinds in the migration, and because so many of them
were teachers they introduced in America, through generations of students,
European intellectual traditions, methods, and points of view. And last, but
certainly not least, a considerable number of atomic scientists came to our
shores and we shall find many of them in Los Alamos.

Although the intellectual migration, as I studied it, began in the early
thirties, we, the Fermis, arrived in New York on January 2, 1939, and we
settled in New Jersey. We thought we had settled there for good; we bought
a house because Fermi had a permanent position at Columbia University.
But three years later we moved to Chicago and the Metallurgical Laboratory;
we thought we had moved there for the duration of the war, but then Oppenheimer came and 'sold' us 'Site Y.' I didn't know at that time that its true
name was Los Alamos — but there we went in the summer of 1944. I was the
only wife so far as I know, to arrive at Site Y and spend the first few weeks
there without a husband. Fermi had planned to go with us but at the last
moment he said he had to go to Richland in Washington State and did not
know for how long. The idea of Site Y had caught the imagination of my
children and myself and rather than wait for Fermi we made arrangements
to go there alone.

As we got off the train at Lamy, New Mexico, a soldier came up to me
and said, "Are you Mrs. Farmer?" I said, "Yes, I'm Mrs. Fermi." He frowned
and said, "I was told to call you Mrs. Farmer." That was the name Fermi was
supposed to use when he was traveling. The soldier motioned us to a car and
we drove off. After a long ride through the most spectacular country I had
ever seen, we reached a barbed-wire fence and a gate. We showed the passes
that had been provided for us to a military policeman and proceeded to a
barrack-styled house and the apartment in it that had been assigned to us. I
knew already the size of the windows in that apartment because physicist
John Manley had measured them accurately so that I could buy curtains in
Chicago. About Site Y, I also knew that it was a muddy place where a pair of
hiking boots would be a blessing. And that was more or less the extent of my
knowledge. Only several days after my arrival did I learn that the name of
Site Y was Los Alamos. Fermi was in Hanford, the secret city near Richland
where the piles for the production of plutonium were being built. A pile is
just an atomic reactor, but at that time reactors were called piles because they

were built by *piling up* graphite and uranium. When the construction of the
Hanford piles began, one pile only had operated: the Chicago pile. This was a
small thing, with rudimentary control rods and no shielding. There was a large
change of scale from the Chicago to the Hanford piles. As Fermi would have
put it, they were different animals. So Fermi had been asked to be on hand
for the starting of the first Hanford pile — just in case something would go
wrong. And something *did* go wrong, and Fermi was detained in Hanford.
Only 15 years later, in 1958, did I learn what had happened. The pile had
produced some gases the effect of which was the 'poisoning' of the pile.
I learned this in the following way.

At the end of the war, Fermi brought me a mimeographed copy of the
so-called Smyth Report. It is now published under the title *Atomic Energy
for Military Purposes*, and it is the history of the whole uranium project. The
copy Fermi brought home had been expurgated and was supposed to contain
only declassified information. Fermi said, "You may be interested in reading
this." It was *such* hard reading that I had to read it word by word and ask
questions now and then. At one point I got to a sentence that said, "the
poisoning effect that may have impeded production" at the Hanford site.
So I turned to Fermi and asked what was the meaning of 'poisoning.' He said,
"What? Is that sentence there? It's supposed to have been crossed out! It
would be an interesting story, but it's a secret." Years later, when I was writing *Atoms in the Family*, I asked him, "What about the poisoning of the piles;
is it still a secret?" Fermi said again that it would make such a nice story, and
that I should go ask Libby whether it was secret or not. So I went to Willard
Libby's office in Chicago, and he pulled out a lot of little cards, pushed a
metal rod through them, then agitated this rod and said, "Yes, the story is
still classified. If you want to have it de-classified, write it and send it to the
Atomic Energy Commission asking permission to publish it." But this was not
possible, because Fermi would have never taken the time to write it himself
and if he told me the story and *I* wrote it, there would be a breach of security
laws. I gave up. Finally in 1958, Compton published his book *Atomic Quest*
in which he told the Hanford episode in which one of the fission fragments
absorbed so many neutrons that it 'poisoned' the pile.

Secrecy fell upon the work of the nuclear scientists not long after we, the
Fermis, landed in the United States. Two weeks after we arrived, Bohr came
from Denmark with the news of the discovery of fission. By 1940, Leo
Szilard, one of the many Hungarians in the intellectual migration who, like
all Hungarians, had a great imagination, had proposed, and scientists had
accepted, secrecy for their work. So, by the time I arrived at Los Alamos, I

had already had four years of increasingly noticeable secrecy. At Columbia University one room or two were closed to the public, but I doubt that many people were aware of it. In Chicago the security measures were more noticeable. We had the first experience with fake names: for instance, 'Metallurgical Laboratory' was a fake name, as there were no metallurgists in it. We called it Met Lab for short. In Chicago we had also the first experience with armed guards posted in the hall of the physics building and prohibiting the wives from entering; with bodyguards for top scientists; and pep talks to the wives to keep their tongues in check. Betty Compton, the wife of the Met Lab's director, invited all newcomer wives to a film called 'Next of Kin' which showed how the carelessness of a British officer caused, in the end, the bombing of London. So, we got the message.

In Los Alamos we met with the added feature of isolation from the world as enforced by the fence, the Army, mail censorship and so on. We were not on the map at all. Our address was Box 1663, Santa Fe, which was an office in town; and for added security some people like us, when writing abroad, had to give an address in Chicago. Even the inhabitants of Santa Fe, where we went on frequent shopping expeditions, knew only that all our shopping was delivered in town, at the office of the United States Army Corps of Engineers, although we lived up on the hill and maybe on the hill there was a camp for pregnant WACs or a submarine base. (I'm sure that the idea of the submarine base must have come from Edward Teller, who is probably the Hungarian-born scientist with the greatest imagination — Los Alamos is a terribly arid place).

Arriving in 1944, we were latecomers and although the population never ceased to grow, the key people were already there. A striking feature was the large number of European-born of relatively recent arrival — they were, of course, part of the intellectual migration of the '30s. For us, it was a pleasure to find so many Europeans. There were more friends in Los Alamos than in other places we had ever lived before. Among the Europeans there were old Italian friends like the Segrès and the Rossis; physicists who had visited us in Rome like Bethe and Peierls; and the Tellers with whom we had become very friendly in this country. Teller was intellectually very close to Fermi, and when they put their heads together their brains gave off sparks. One of the sparks was the idea of achieving a thermonuclear reaction on earth. They discussed this idea when we were still in New York at a luncheon at the Faculty Club at Columbia. And from then on Teller kept on thinking about thermonuclear explosions. There were others with whom we became friends at Los Alamos, like the Ulams and the Staubs with whom we shared an affinity of background. And there was of course 'Uncle Nick' who had been very kind to us when we had stopped in Copenhagen on our way to the United States and whom we couldn't call Niels Bohr or Professor Bohr because it was a most top secret that Bohr was on the mesa.

These foreigners, each with his own peculiarities, were colorful and stuck out. They were a minority in number but a very influential minority. All were brilliant, all refugees from the totalitarian countries. They were brilliant because strong selective forces had acted on them and made it possible for them to come to this country. In retrospect, the phenomenon appears amazing. Think of the paradox. No other country but the United States would have entrusted foreigners, some of them from enemy countries, with work in the most secret project, that involved the national defense. Many of these scientists technically became enemy aliens when America entered the war. They had applied for citizenship, but it takes five years to become a citizen, so some of the people at Los Alamos weren't even United States citizens. Was our government taking undue risks?

Secrecy had strong indirect effects on life in Los Alamos. Tensions were plentiful. When I arrived in Los Alamos my old friends Nora Rossi, Rose Bethe and Mici Teller came to me, one at a time, to tell me their grudges. Mici Teller, for instance, didn't see why the kindergarten people wouldn't take her baby because he wasn't trained. Another woman complained that somebody else was attempting to take a lead in the local social life while she herself was much better qualified for this role. And when Fermi finally arrived, at least three physicists, and possibly more, came up to me and said they wanted to be the first ones to talk to him. I didn't know, of course, what they wanted to talk about, but it was clear that they were under tension, wanted to clarify some situation and hoped for Fermi's help.

The tensions were caused by several factors. We were too many of one kind, all packed together. We could not unburden our souls by bringing our complaints to outsiders; they had to remain either within ourselves or within the community. None of us had lived in such conditions, conditions so similar to those of a concentration camp that some European-born scientists could not stand them and quit. It was not only the barbed-wire fence, the passes and the badges, and the inspections of cars going in and out of town. It was also the Army running the town uncontested and on a socialist basis. The apartments, for instance, were assigned to us according to the number of persons in the family. We had no voice in choosing them and we paid rent according to salaries! The machinist downstairs in our own building probably paid half the rent we did. Our doctors' services were free. Household help was available but rationed. In the morning, Army buses rounded up Indian and Spanish-American girls in their pueblos and villages and brought them to the mesa. In the beginning the help was plentiful, but as the technical area absorbed more people, and as the population of Los Alamos grew, the help became inadequate and so it was rationed according to complicated rules that took into account pregnancies, the number of children, number of hours a wife was working, and so on. For the logistic operations, such as assigning apartments, the Army delegated its power to the housing office which was run first by wives and then by WACs.

All this created resentments and there were grumblings. We felt that the the women in charge of apartments and those in charge of maids were too inexperienced and prone to favoritism. For instance, when a person as touchy as Emilio Segrè was assigned, without possibility of recourse, an apartment above the band leader, who practiced at home with the full band, he could not but think that the wife in charge of apartments had intentionally snubbed him. There was even a rumor that the woman in charge of counseling tried to dissuade a physicist from getting married rather than try to find an apartment for him. Even the free medical services were a reason for discontent. We felt that they turned out to be as undemocratic as many other things in Los Alamos. Our Army doctors were very nice, and I felt sorry for their thankless job. They had prepared to treat the wounded on the battlefield, and instead they were faced with a bunch of high-strung, healthy civilians with minor ailments and an incredible number of babies, already born or to be delivered! When hospitalized, we paid only for food, so the standard cost of a baby was fourteen dollars. I think that young women couldn't resist the temptation of such a bargain! (Fortunately I was a little older.) But of course we could not choose our doctor so we criticized those residents who took up all the doctor's time, we thought, with imaginary complaints, just because the services were free; and we criticized as well those who grumbled that unpaid doctors could *not* give enough attention to them and their kids.

The complaints against the Army were as numerous and even more futile than all other complaints put together. The fingerprinting of civilians; the bureaucracy of filling forms in many copies, even for the replacement of a light bulb; the loop holes in gas rationing; all were causes of irritation. But, against the Army we were powerless. A city council had been organized to channel complaints to the high powers, but to my knowledge it was not very effective. And so we kept grumbling.

These tensions and irritations were the price, minor in retrospect, that we had to pay so that the project could go ahead in secrecy and achieve what it did. The effects of secrecy within single families were more difficult to assess. There were great differences in the ways wives were treated. I've been accused of greatly exaggerating when, in *Atoms in the Family*, I wrote that throughout the war I didn't know anything about the project. But I know there were other wives who didn't know anything. On the other hand, several husbands felt that in view of the isolation of Los Alamos and the security measures taken there, it was safe to discuss a number of matters with their wives. So far as I know, these wives never leaked any information to the others.

Fermi was extremely tight-lipped. Once in Chicago when he was at the Met Lab, I related to him the rumour that the scientists at Met Lab were working toward a cure for cancer. Fermi assumed his best poker face and said, "Do we?," and I said, "So you don't?' He replied, "Don't we?" So there wasn't much point in trying to pump information from him.

I may give you an idea of what secrecy did to the Fermi family, and I shall start from the youngest member, my son Giulio who was two when we left Italy. In this country, never hearing anything about his father's work, he underrated his father's role until Hiroshima and the explosions of comments that followed. Until then, Fermi was not an important man to Giulio. At least, not as important as George's father who was a Captain in the Army, as Giulio once said. Los Alamos gave a boost to his ego. Giulio enjoyed playing the secrecy game, as indeed most of us did, up to a certain extent. And he was among the kids who showed adults how to get out of the site through the holes in the fence that they had discovered. He enjoyed cheating the guard at the gate by lying low in the car, pretending not to be there, and thus confusing the record of who went out of the site and who came back. Giulio himself never looked like the son of an important man. He went to camp that summer of 1945, and was asked, "Are you any relation to Dr. Fermi?" When he answered, "I'm his son," he was told, "Impossible." Something similar happened to me the first time I returned to Europe. I had a suit made by a modest tailor in the mountains and he asked me, "Are you any relation to Dr. Fermi?" I said, "I'm his wife," and he said, "Impossible, he is a millionaire!" What wife of an important man would have a suit made by such a modest tailor?

My daughter Nella, on the other hand, was seven when on our way to the United States we stopped in Sweden to let Fermi pick up his Nobel Prize. She attended the Nobel ceremony, and I don't think that in her case her father's importance was ever an issue. Besides, she had her own difficulties. She changed too many schools and had to adapt to too many places. In Los Alamos she was thirteen and happy. Her teachers at the site were Los Alamos wives, I mean scientists' wives: Alice Smith (Mrs. Cyril Smith) and Jane Wilson (Mrs. Robert Wilson) well established in the secrecy pattern and ready to give her recognition for her talents. Once in class, Nella and her teacher Jane Wilson had an argument about the structure of the atom. Nella was right, Jane wrong, and Jane promptly admitted it. Nella, however, had a traumatic experience. When Fermi was not traveling, his bodyguard and friend, John Baudino, was employed in the security office where our mail was censored. Once Baudino happened to read a letter from Nella to a friend, enjoyed it and talked about it to Fermi. Nella was so mad at this invasion of privacy that she never forgave it, and it may well be at the roots of her present stand as an exponent of women's lib.

As for myself, I greatly enjoyed my vacation from physics. Only physicists think that physics *must* be the main topic of conversation in a family. I loved Los Alamos, though it kept me very busy. We had no telephones and we ran around a lot. We gave large parties, cooking on rudimentary appliances. We rushed to Santa Fe for anything that was not food or the little else that the Army could sell. After the liberation of Rome, we went on errands for Italians who asked for all sorts of things and who once sent even a little piece of printed cloth asking me to match it. They didn't realize we were in Los Alamos, of course; because of the return address on our letters they thought we were in Chicago.

I had a part-time job in the technical area, working 3/8ths of the time as a blue badge, who could not be told any secrets. (People who could be told secrets were white badges − physicists and chemists and so on.) The Army encouraged wives to work 'to keep them out of mischief.' I was in charge of the personnel files in the doctor's office and I knew who was shifted from one part of the project to the other, where everybody lived, whose blood count was low, and other trivia. I passed on this information to Fermi. He was assistant director of the project and was also the head of a division that was nicknamed 'Problem Division,' because it solved problems that other divisions couldn't solve − and also was made up of problem children. (One of them was Edward Teller, who insisted on working on the thermonuclear reaction rather than on the atomic bomb.) Anyhow, there were problem children even in Los Alamos. However, despite his position, Fermi never seemed to know what I did and since he never talked about his work, I was under the impression that I knew everything about the project and he knew nothing; it was a very good feeling that I owed to secrecy.

I feel that, despite secrecy, I should not have been as unprepared for the news of the bomb as I was. I had seen the origin of the work that led to it in 1934 in Rome when Fermi and Segrè and two other physicists started to bombard with neutrons all elements they could get. Neutrons were at that time only two years old, since Chadwick had discovered them in 1932. And artificial radioactivity had just been produced for the first time in a small number of elements by the Joliot-Curies, who used alpha particles. Fermi decided to try neutrons. Geiger counters were not standard laboratory equipment in 1934, so Fermi had to build his own. With the help of his friends, he also prepared neutron sources made of radon and beryllium. Segrè was what now would be called procurement. He went with a shopping list and a shopping bag to visit all chemists' shops in Rome, inspecting their dustiest shelves, and he even borrowed some gold from a jeweler friend. The group began bombarding the elements systematically in the order of the periodic table. The lightest elements did not react, but most others, from fluorine up, became radioactive. The last element they bombarded was, of course, uranium and it behaved very strangely. An accurate separation of the products obtained would have been almost impossible at that time as the amounts of activated substances were extremely small. Besides, they were physicists. They had in their group one chemist, but the techniques were not well developed. The physicists suspected that they had created a new element, 93, but they realized that before being sure they would have to make many more tests. At this point something unexpected diverted their attention. In the fall of the year 1934, they accidentally discovered slow neutrons and the fact that these could produce as much as 100 times more radioactivity than fast neutrons. So Fermi gave himself to this new line of research.

I had followed Fermi's work from behind the scenes and was aware of most of its aspects. Four years later, when Bohr introduced the word fission in this country, Fermi explained to me that Otto Hahn, Fritz Strassmann, and Lise Meitner in Germany had picked up the uranium experiment where the Roman group had left it. And after four years of patient research and uncertain results, they had come up with a correct interpretation. When the Roman group had bombarded uranium in 1934, they had not produced element 93. They had produced fission but failed to recognize it. (That was the situation in 1939. A few years later, as the process was explained even further, it became evident that Fermi and his friends had produced also a few atoms of 93, which is now called neptunium.) Well, Fermi told me that fission might be the key that could unlock the huge stores of energy in atoms. If so, fission might be of military interest, for instance, to produce atomic energy to drive battleships. Explosions were not mentioned in my presence, with two exceptions. Sometime in 1939 or '40 Mici Teller lent me a book urging me to read it: it was the story of an international incident caused by the explosion of an atomic bomb. Later, when we were in Chicago, Segrè once came to visit. He stayed with us and also went to the laboratories with Enrico to see what he was doing. Then Segrè came back and said, "If Fermi blows up, you'll blow up too." But he has always been a pessimist foreseeing dramatic accidents, and I paid no attention to him. These two mentions of explosions made very little impression on me. Had I done more thinking over what I knew I might have guessed a bit more, but not too much. Mental inertia, encouraged by the fact that there was no point in trying to pump information out of Fermi, and involvement in my daily, seemingly pressing occupations are my only excuses.

You may ask: what were these pressing occupations? I'll give you one example. Occasionally, the news ran on the mesa that a shipment of Army turkeys had arrived at the commissary store. At that time meat was rationed but poultry was not, so the arrival of turkeys meant that we could give a party. We rushed to the commissary. The Army turkeys, no less than 22 to 25 pounds, were frozen but not eviscerated. The butcher was very kind and he'd say, "Bring it back when it's thawed and I'll draw it for you." But of course I'd have to lug back some 25 pounds. So I learned how to draw a turkey from Segrè, who being a physicist, told me how many pieces I should recover from the turkey. He also warned me to be careful not to break the bag containing the bile. A more serious occupation was sending food packages to Italy. I remember my qualms of conscience when we were allotted additional sugar for canning and instead of canning I sent the sugar to Italy. I felt that in a sense I was cheating. It was a moral decision, a small one, but still a moral decision to make.

When someone is fully involved in daily occupations, he's likely to neglect lines of thought which hindsight tells him he should have followed. The Los Alamos scientists themselves are a signal example. They were so involved with their work and under such pressure of time that they gave little thought to what later became known as the "social implications of the bomb." Of course they would not have discussed these matters at home, but I have reason to believe that they did *not* discuss them at work either. The reason lies in a meeting we had with a writer, Dexter Masters, who came to Los Alamos a few years after the war to write a novel on the Slotin accident. Louis Slotin was one of two casualties at Los Alamos. One happened in '45; a young man named Harry Daghlian broke safety rules and went alone, at night, to work in a laboratory. Something went wrong, he grabbed some fissionable material with his hands, to pull the pieces apart and stop the chain reaction, but died of radiation a couple of weeks later. The next year (we were no longer in Los Alamos) Slotin was showing a group of physicists a critical assembly made of two pieces. He was keeping the two pieces apart with a screwdriver. Suddenly the screwdriver slipped, the two pieces came in contact, and the assembly became critical. Slotin, being the closest to the assembly received the largest amount of radiation and eventually died. Graves who was the next closest to the assembly, temporarily became completely bald, and developed other symptoms, including cataracts, due to exposure to neutrons.

Dexter Masters, trying to recreate the background to Slotin's accident, wanted to know what we were talking about during the war. We searched our memories — We talked, we said, about smuggling liquor into Los Alamos, about the road that was too steep and winding, about the WACs who drove too fast, and so on. But that was not what he wanted, so he finally asked us directly whether we ever talked about the social implications of the bomb. Priscilla Duffield put the matter in a nutshell. She had been Oppenheimer's secretary, she said, and she would have known if any discussions of that kind had taken place. But there were none.

Now Priscilla Duffield's assertion needs qualification. A considerable amount of discussion had gone on at the Met Lab in Chicago where the pressure of work was not as great as in Los Alamos at that moment. And the best known document coming out of these discussions was the Franck Report. Szilard also prepared a petition asking the President not to use the bomb. According to Teller in his book *The Legacy of Hiroshima*, Szilard asked Teller to circulate the petition in Los Alamos. Teller consulted Oppy, and Oppy told him (I quote Teller) that he thought it "improper for a scientist to use his prestige as a platform for political pronouncements." Oppy felt that the decision was in the hands of the best, the most conscientious and wise men in the nation and the scientists did not need to worry. The petition was not circulated. There may have been other discussions, more secret, too secret even for the office of the director: In late spring, 1945, four scientists, Fermi, Oppenheimer, Compton and Lawrence, two of whom were in Los Alamos, were appointed, as a scientific panel, to advise the Interim Committee on matters related to the atomic bomb. They submitted a report that has been published many years later and can be found in Alice Smith's book, *A Peril and a Hope*. Since Oppy and Fermi were in Los Alamos, they had probably exchanged views about the use of the bomb.

As a whole, however, the Los Alamos scientists had not given much thought to the social implications of the atomic bomb. This explains the

outburst of words, feelings, emotions and expressions of a sense of guilt in
Los Alamos right after Hiroshima. Suddenly, to the amazement of us wives,
our husbands talked of nothing else but the bomb. They seemed to carry on
their shoulders the responsibility for the havoc done in Japan and for the evils
that atomic energy might bring anywhere in the world in the future. One
sentence stands out in my memory because I did *not* fully understand it at
the time, and felt it might be due to that arrogance of which physicists have
occasionally been accused: "We must educate the public." Why should *they*
educate the public? But of course, they were right. Unless the public came to
understand some of the basic scientific, technological and political aspects of
the bomb, the Americans would not be able to make well-reasoned decisions.
Our husbands went ahead and gave lectures in Santa Fe and Albuquerque and
they organized the Association of Los Alamos Scientists. They drafted state-
ments, they wrote articles.

All this was much harder to take than secrecy. Not only had I to read and
digest, without much help from Fermi, the Smyth Report, which gave me
merely five years' worth of history of physics, I had also to absorb the philos-
ophy of atomic energy, to grasp the meaning of those social implications of
which there was now so much talk. Should I side with those who wanted
America to share the secrets of the bomb, whatever these were, with all countries including Russia, immediately, just to show our good will? Or should I
side with those who felt that as long as some nations were protecting themselves with an impenetrable iron curtain, the world was not yet ready to have
and administer atomic energy? These were hard decisions to make. But above
all, there were the moral questions. I knew scientists had hoped that the
bomb would not be possible, but there it was and it had already killed and
destroyed so much. Was war or was science to be blamed? Should the scientists have stopped the work once they realized that a bomb was feasible?
Could they have stopped it?. Would there always be war in the future? To
these kinds of questions there is no simple answer.

I would like to ask again, did our government take undue risks when it
allowed recent emigrés to work at one of its most important wartime projects?
As you know, there was one spy, Klaus Fuchs, and through him information
leaked to Russia. It was evaluated that probably because of Fuchs, Russia
made the atomic bomb some few years earlier than she would have otherwise.
He was a member of the British Mission, not a refugee in America. All other
foreign-born were as devoted to America as American-born, as determined if
not more determined not to let the dictators win the war. Not all the foreign-
born scientists in the Manhattan Project were in Los Alamos. Szilard and

James Franck were in Chicago. Wigner spent most of his time at Site X, Oak Ridge, that is. Von Neumann and Bohr came to Los Alamos only on visits.

All these scientists contributed to the success of the project. But you may ask, to what extent did they contribute? I put this question to many scientists, both Americans and European-born, and they seemed to agree on two points. One, the Europeans had a decisive role in the development of atomic energy. Without their presence the release of this energy would have been delayed by many years and atomic energy would not have been produced before the end of the war. Had the Europeans stayed in their countries they would not have achieved the release of atomic energy abroad. The odds were enormous and even Germany, which was widely regarded as the most advanced country, scientifically, had given up. But of course, we did not know that Germany had given up. For success several factors were needed! An *enormous* concentration of brain power in one country; the very close collaboration of European and American scientists with their different skills and intellectual traits; their unity in their will to defeat the dictators; the formidable American industrial and financial power; and, we must admit in spite of all our wartime resentment, the coordinating supervision of the Army.

The friends I talked to agreed also on another point that may somewhat sober the achievers. The most crucial contribution to victory in World War II was radar, not atomic energy. Radar won the war, atomic energy shortened it. In the radar development, the Europeans had only a limited role. Felix Bloch, a European, collaborated in that project and so did Sam Goudsmit, a Dutch-born physicist. All considered, we may conclude that the government's trust in the refugees to whom it had given asylum was not misplaced. It was a reason for great gratitude that may have enhanced their loyalty.

DISCUSSION

Question: Were any of the women at Los Alamos dissatisfied that they couldn't play a role in the scientific endeavor or *were* there some that were involved?

Fermi: There *were* several; for instance, Joan Hinton, one of many young people who interrupted college to come to Los Alamos. After Los Alamos, she came to Chicago to get her master's. But she never finished her studies because she went to China to join her fiancé and get married. You may have read about her in the papers. She was a great admirer of Chinese Communists for all they did to help the little people in small villages.

Mary Argo was a physicist and did have some responsibility in the project;

but she had also to take care of her children. I don't think there were resentments. I think that the project tried to make use of the women's abilities whenever possible. There were women doctors and biologists mainly from St. Louis. There was also Mrs. Graves, a physicist. So I think that to the extent of their abilities, women did participate in the project. Quite a few of them were teachers as I indicated.

Question: Would you say some more about the commune-nature of life in Los Alamos?
Fermi: As long as I was in Los Alamos, in wartime and during the following summers I went back, there was this kind of living, the closest I have seen in America to a communistic type of living on a somewhat large scale. We were something like 5,000 people by the end of the war. I haven't seen similar conditions in any other place. Much later the Atomic Energy Commission started selling the houses to residents and the old spirit disappeared, but I am not really able to put a dividing line between the old and new ways of living. One interesting thing to me is that during the war I was in Los Alamos only a year and a half or even less, and *still* it seems such a big portion of my life. I don't know if I can convey what I mean. It was so different, it was such intense living, that it seems impossible that in my case it actually lasted hardly eighteen months.

Question: How critical was the timing of your emigration from Italy in 1939?
Fermi: Critical in what respect? An interesting fact emerged when I was talking to emigrés for a book I wrote. I talked to quite a few of them and I usually asked them whether they were forced out of their countries. Invariably, they did not consider that they had been *forced* out; they had made their own decisions, they said, and used their free wills. They were not kicked out, and could have stayed. Maybe they would have been killed, but still there was a choice. In Italy things got really bad for Jewish people, of which I am one, when the Germans invaded Italy. The Fascists had compiled lists of Jews and promulgated pretty strict laws, some of which were enforced. All Jews lost their positions, among them Segrè and Rossi. Fermi was not a Jew so he could have stayed if he wanted to. Our children could have gone to public school because they were children of a Jew and a non-Jew and had been baptised. You see what kind of illogical divisions there were; I mean, what kind of thinking was behind the laws? Our children could have gone to public schools, but children whose parents were both Jews could not. These

laws were put into effect right away, at the end of '39. On the other hand, there was no danger to life until '43, when the Germans invaded Italy.

Question: Did Fermi believe that scientists should organize into groups such as the Federation of American Scientists for political purposes?
Fermi: I think that at that point he felt that the world was not ready for the aims of that Federation. In general, he felt that as a man of science, he was more productive and useful as long as he did science than when collaborating in political activities. On the other hand, he was a co-writer of one of the reports that came out of the Met Lab, the *Report on Nucleonics*, on the future of atomic energy. After the war he collaborated with the Atomic Energy Commission as a member of the General Advisory Committee. So, he did his share, but he still felt that his major contributions could be in science rather than in politics.

Question: Do you think that secrecy was carried too far? Were you aware of events occurring outside of Los Alamos?
Fermi: We kept informed of what was going on in the outside world. We had the radio and the newspapers, and we talked about the war. I remember several discussions. I remember, for instance, very vividly, Bohr, on a hike down a canyon. He kept talking about the war in Europe, the doom of Europe and his own experiences in Denmark. We were alert and we were talking about the general situation. There was, however, very little knowledge of what other parts of the project were doing.

RICHARD P. FEYNMAN

LOS ALAMOS FROM BELOW

When I say "Los Alamos From Below," I mean that, although in my field at the present time I'm a slightly famous man, at that time I was not anybody famous at all. I didn't even have a degree when I started to work with the Manhattan Project. Many of the other people who tell you about Los Alamos — people in higher echelons — worried about some big decisions. I worried about no big decisions. I was always flittering about underneath.

So I want you to just imagine this young graduate student that hasn't got his degree yet but is working on his thesis, and I'll start by saying how I got into the project, and then what happened to me.

I was working in my room at Princeton one day when Bob Wilson came in and said that he had been funded to do a job that was a secret, and he wasn't supposed to tell anybody, but he was going to tell me because he knew that as soon as I knew what he was going to do, I'd see that I had to go along with it. So he told me about the problem of separating different isotopes of uranium to ultimately make a bomb. He had a process for separating the isotopes of uranium (different from the one which was ultimately used) that he wanted to try to develop. He told me about it, and he said, "There's a meeting —."

I said I didn't want to do it.

He said, "All right, there's a meeting at three o'clock. I'll see you there."

I said, "It's all right that you told me the secret because I'm not going to tell anybody, but I'm not going to do it."

So I went back to work on my thesis — for about three minutes. Then I began to pace the floor and think about this thing. The Germans had Hitler and the possibility of developing an atomic bomb was obvious, and the possibility that they would develop it before we did was very much of a fright. So I decided to go to the meeting at three o'clock.

By four o'clock I already had a desk in a room and was trying to calculate whether this particular method was limited by the total amount of current that you get in an ion beam, and so on. I won't go into the details. But I had a desk, and I had paper, and I was working as hard as I could and as fast as I could, so the fellows who were building the apparatus could do the experiment right there.

It was like those moving pictures where you see a piece of equipment go

105

L. Badash, J. O. Hirschfelder and H. P. Broida (eds.), *Reminiscences of Los Alamos 1943–1945, 105–132.*
Copyright © 1980 by D. Reidel Publishing Company.

bruuuuup, bruuuuup, bruuuuup. Every time I'd look up, the thing was getting bigger. What was happening, of course, was that all the boys had decided to work on this and to stop their research in science. All science stopped during the war except the little bit that was done at Los Alamos. And that was not much science; it was mostly engineering.

All the equipment from different research projects was being put together to make the new apparatus to do the experiment — to try to separate the isotopes of uranium. I stopped my own work for the same reason, though I did take a six-week vacation after a while and finished writing my thesis. And I did get my degree just before I got to Los Alamos — so I wasn't quite as far down the scale as I led you to believe.

One of the first interesting experiences I had in this project at Princeton was meeting great men. I had never met very many great men before. But there was an evaluation committee that had to try to help us along, and help us ultimately decide which way we were going to separate the uranium. This committee had men like Compton and Tolman and Smyth and Urey and Rabi and Oppenheimer on it. I would sit in because I understood the theory of the process of what we were doing, and so they'd ask me questions and talk about it. In these discussions one man would make a point. Then Compton, for example, would explain a different point of view. He would say it should be *this* way, and he would be perfectly right. Another guy would say, well, maybe, but there's this other possibility we have to consider against it.

I'm jumping! Compton should say it *again!* So everybody is disagreeing, all around the table. Finally, at the end, Tolman, who's the chairman, would say, "Well, having heard all these arguments, I guess it's true that Compton's argument is the best of all, and now we have to go ahead."

It was such a shock to me to see that a committee of men could present a whole lot of ideas, each one thinking of a new facet, while remembering what the other fellow said, so that, at the end, the decision is made as to which idea was the best — summing it all up — without having to say it three times. So that was a shock. These were very great men indeed.

It was ultimately decided that this project was *not* to be the one they were going to use to separate uranium. We were told then that we were going to stop, because in Los Alamos, New Mexico, they would be starting the project that would actually make the bomb. We would all go out there to make it. There would be experiments that we would have to do, and theoretical work to do. I was in the theoretical work. All the rest of the fellows were in experimental work.

The question was — What to do now? Los Alamos wasn't ready yet. Bob

Wilson tried to make use of this time by, among other things, sending me to Chicago to find out all that we could find out about the bomb and the problems. Then, in our laboratories, we could start to build equipment, counters of various kinds, and so on, that would be useful when we got to Los Alamos. So no time was wasted.

I was sent to Chicago with the instructions to go to each group, tell them I was going to work with them, and have them tell me about a problem in enough detail that I could actually sit down and start to work on it. As soon as I got that far, I was to go to another guy and ask for another problem. That way I would understand the details of everything.

It was a very good idea, but my conscience bothered me a little bit because they would all work so hard to explain things to me, and I'd go away without helping them. But I was very lucky. When one of the guys was explaining a problem, I said, "Why don't you do it that way?" In half an hour he had it solved, and they'd been working on it for three months. So, I did something! Then I came back from Chicago, and I described the situation — how much energy was released, what the bomb was going to be like, and so forth.

I remember a friend of mine who worked with me, Paul Olum, a mathematician, came up to me afterwards and said, "When they make a moving picture about this, they'll have the guy coming back from Chicago to make his report to the Princeton men about the bomb. He'll be wearing a suit and carrying a briefcase and so on — and here you're in dirty shirtsleeves and just telling us all about it, in spite of its being such a serious and dramatic thing."

There still seemed to be a delay, and Wilson went to Los Alamos to find out what was holding things up. When he got there, he found that the construction company was working very hard and had finished the theater, and a few other buildings that they understood, but they hadn't gotten instructions clear on how to build a laboratory — how many pipes for gas, how much for water. So Wilson simply stood around and decided, then and there, how much water, how much gas, and so on, and told them to start building the laboratories.

When he came back to us, we were all ready to go and we were getting impatient. So they all got together and decided we'd go out there anyway, even though it wasn't ready.

We were recruited, by the way, by Oppenheimer and other people, and he was very patient. He paid attention to everybody's problems. He worried about my wife who had TB, and whether there would be a hospital out there, and everything. It was the first time I met him in such a personal way; he was a wonderful man.

We were told to be very careful – not to buy our train ticket in Princeton, for example, because Princeton was a very small station, and if everybody bought train tickets to Albuquerque, New Mexico, in Princeton, there would be some suspicions that something was up. And so everybody bought their tickets somewhere else, except me, because I figured if everybody bought their tickets somewhere else . . .

So when I went to the train station and said, "I want to go to Albuquerque, New Mexico," the man says, "Oh, so all this stuff is for *you!*" We had been shipping out crates full of counters for weeks and expecting that they didn't notice the address was Albuquerque. So at least I explained why it was that we were shipping all those crates; *I* was going out to Albuquerque.

Well, when we arrived, the houses and dormitories and things like that were not ready. In fact, even the laboratories weren't quite ready. We were pushing them by coming down ahead of time. So they just went crazy and rented ranch houses all around the neighborhood. We stayed at first in a ranch house and would drive in in the morning. The first morning I drove in was tremendously impressive. The beauty of the scenery, for a person from the East who didn't travel much, was sensational. There are the great cliffs that you've probably seen in pictures. You'd come up from below and be very surprised to see this high mesa. The most impressive thing to me was that, as I was going up, I said that maybe there had been Indians living here, and the guy who was driving stopped the car and walked around the corner and pointed out some Indian caves that you could inspect. It was very exciting.

When I got to the site the first time, I saw there was a technical area that was supposed to have a fence around it ultimately, but it was still open. Then there was supposed to be a town, and then a *big* fence further out, around the town. But they were still building, and my friend Paul Olum, who was my assistant, was standing at the gate with a clipboard, checking the trucks coming in and out and telling them which way to go to deliver the materials in different places.

When I went into the laboratory, I would meet men I had heard of by seeing their papers in the *Physical Review* and so on. I had never met them before. "This is John Williams," they'd say. Then a guy stands up from a desk that is covered with blueprints, his sleeves all rolled up, and he's calling out the windows, ordering trucks and things going in different directions with building material. In other words, the experimental physicists had nothing to do until their buildings and apparatus were ready, so they just built the buildings – or assisted in building the buildings.

The theoretical physicists, on the other hand, could start working right away, so it was decided that they wouldn't live in the ranch houses, but would live up at the site. We started working immediately. There were no black-boards except for one on wheels, and we'd roll it around and Robert Serber would explain to us all the things that they'd thought of in Berkeley about the atomic bomb, and nuclear physics, and all these things. I didn't know very much about it; I had been doing other kinds of things. So I had to do an awful lot of work.

Every day I would study and read, study and read. It was a very hectic time. But I had some luck. All the big shots except for Hans Bethe happened to be away at the time, and what Bethe needed was someone to talk to, to push his ideas against. Well, he comes in to this little squirt in an office and starts to argue, explaining his idea. I say, "No, no, you're crazy. It'll go like this." And he says, "Just a moment," and explains how *he's* not crazy, *I'm* crazy. And we keep on going like this. You see, when I hear about physics, I just think about physics, and I don't know who I'm talking to, so I say dopey things like, "No, no, you're wrong," or "You're crazy." But it turned out that's exactly what he needed. I got a notch up on account of that, and I ended up as a group leader under Bethe with four guys under me.

I had a lot of interesting experiences with Bethe. The first day when he came in, we had a calculator, or glorified adding machine, a Marchant that you work by hand. And so he said, "Let's see." The formula he'd been working out, he says, "involves the pressure squared; the pressure is 48; so the square of 48 is —."

I reach for the machine.

He says, "It's about 2300." So I plug it out just to find out.

He says, "You want to know exactly? It's 2304." And it came out 2304.

So I said, "How do you do that?"

He says, "Don't you know how to take squares of numbers near 50? If it's near 50, say 3 below (47), then the answer is 3 below 25 — like 47 squared is 2200, and how much is left over is the square of what's residual. For instance, it's 3 less and the square of that is 9, so you get 2209 from 47 squared."

So he knew all his arithmetic, and he was very good at it, and that was a challenge to me. I kept practicing. We used to have a little contest. Every time we'd have to calculate anything we'd race to the answer, he and I, and I would lose. After several years I began to get in there once in a while, maybe one out of four. You have to *notice* the numbers, you see — and each of us would notice a different way. We had lots of fun.

Well, when I was first there, as I said, the dormitories weren't ready. But the theoretical physicists had to stay up there anyway. The first place they put us was in an old school building — a boys' school that had been there previously. I lived in a thing called the Mechanics' Lodge. We were all jammed in there in bunk beds, and it wasn't organized very well because Bob Christy and his wife had to go to the bathroom through our bedroom. So that was very uncomfortable.

The next place we moved to was called the Big House, which had a balcony all the way around the outside on the second floor, where all the beds were lined up next to each other, along the wall. Downstairs there was a big chart that told you what your bed number was and which bathroom to change your clothes in. Under my name it said "Bathroom C," but no bed number! By this time I was getting annoyed.

At last the dormitory was built. I went down to the place where rooms were assigned, and they said, you can pick your room now. You know what I did? I looked to see where the girls' dormitory was, and then I picked a room that looked right across — though later I discovered a big tree was growing right in front of the window of that room.

They told me there would be two people in a room, but that would only be temporary. Every two rooms would share a bathroom, and there would be doubledecker bunks in each room. But I didn't *want* two people in the room.

The night I got there, nobody else was there, and I decided to try to keep my room to myself. Now my wife was sick with TB in Albuquerque, but I had some boxes of stuff of hers. So I took out a little nightgown, opened the top bed, and threw the nightgown carelessly on it. I took out some slippers, and I threw some powder on the floor in the bathroom. I just made it look like somebody else was there. OK? So, what happened? Well, it's supposed to be a men's dormitory, see? So I came home that night, and my pajamas are folded nicely, and put under the pillow at the bottom, and my slippers put nicely at the bottom of the bed. The lady's nightgown is nicely folded under the pillow, the bed is all fixed up and made, and the slippers are put down nicely. The powder is cleaned from the bathroom and *nobody* is sleeping in the upper bed.

Next night, the same thing. When I wake up, I rumple up the top bed, I throw the nightgown on it sloppily and scatter the powder in the bathroom and so on. I went on like this for four nights until everybody was settled and there was no more danger that they would put a second person in the room. Each night, everything was set out very neatly, even though it was a men's dormitory.

I didn't know it then, but this little ruse got me involved in politics. There were all kinds of factions there, of course — the housewives faction, the mechanics faction, the technical peoples faction, and so on. Well, the bachelors and bachelor girls who lived in the dormitory felt they had to have a faction too, because a new rule had been promulgated: No Women in the Men's Dorm. Well, this is absolutely ridiculous! After all, we are grown people! What kind of nonsense is this? We had to have political action. So we debated this stuff, and I was elected to represent the dormitory people in the Town Council.

After I'd been in it for about a year and a half, I was talking to Hans Bethe about something. He was on the big Governing Council all this time, and I told him about this trick with my wife's nightgown and bedroom slippers. He started to laugh. "So *that's* how you got on the Town Council," he says.

It turned out that what happened was this. The woman who cleans the rooms in the dormitory opens this door, and all of a sudden there is trouble: Somebody is sleeping with one of the guys! Shaking, she doesn't know what to do. She reports to the chief charwoman, the chief charwoman reports to the lieutenant, the lieutenant reports to the major. It goes all the way up, through the generals to the Governing Board.

What are they going to do? What are they going to do? They're going to think about it, that's what! But, in the meantime, what instructions go down through the majors, down through the captains, through the lieutenants, through the chars' chief, through the char-woman? "Just put things back the way they are, clean 'em up, and see what happens." OK? Next day, same report. For four days, they worried up there about what they're going to do. Finally they promulgated a rule: No Women in the Men's Dormitory! And that caused such a *stink* down below that they had to elect somebody to represent the . . . . . .

I would like to tell you something about the censorship that we had there. They decided to do something utterly illegal and censor the mail of people inside the United States — which they have no right to do. So it had to set up very delicately as a voluntary thing. We would all volunteer not to seal the envelopes of the letters we sent out, and it would be all right for them to open letters coming in to us; that was voluntarily accepted by us. We would leave our letters open; and they would seal them if they were OK. If they weren't OK in their opinion, they would send the letter back to us with a note that there was a violation of such and such a paragraph of our "understanding."

So, very delicately amongst all these liberal-minded scientific guys, we finally got the censorship set up, with many rules. We were allowed to comment on the character of the administration if we wanted to, so we could write our senator and tell him we don't like the way things are run, and things like that. They said they would notify us if there were any difficulties.

So it was all set up, and here comes the first day for censorship: Telephone! *Briiing!*

Me: "What?"

"Please come down."

I come down.

"What's this?"

"It's a letter from my father."

"Well, what is it?"

There's lined paper, and there's these lines going out with dots — four dots under, one dot above, two dots under, one dot above, dot under dot . . .

"What's that?"

I said, "It's a code."

They said, "Yes, it's a code, but what does it say?"

I said, "I don't know what it says."

They said, "Well, what's the key to the code? How do you decipher it?"

I said, "Well, I don't know."

Then they said, "What's this?"

I said, "It's a letter from my wife — it says TJXYWZ TW1X3."

"What's that?"

I said, "Another code."

"What's the key to it?"

"I don't know."

They said, "You're receiving codes, and you don't know the key?"

I said, "Precisely. I have a game. I challenge them to send me a code that I can't decipher, see? So they're making up codes at the other end, and they're sending them in, and they're not going to tell me what the key is."

Now one of the rules of the censorship was that they weren't going to disturb anything that you would ordinarily do, in the mail. So they said, "Well, you're going to have to tell them please to send the key in with the code."

I said, "I don't *want* to see the key!"

They said, "Well, all right, we'll take the key out."

So we had that arrangement. OK? All right. Next day I get a letter from my wife that says, "It's very difficult writing because I feel that the —— is looking over my shoulder." And where the word was, there is a splotch made with ink eradicator.

So I went down to the bureau, and I said, "You're not supposed to touch the incoming mail if you don't like it. You can look at it, but you're not supposed to take anything out."

They said, "Don't be ridiculous. Do you think that's the way censors work —— with ink eradicator? They cut things out with scissors."

I said OK. So I wrote a letter back to my wife and said, "Did you use ink radicator in your letter?" She wrote back, "No, I didn't use ink eradicator in my letter, it must have been the —— " – and there's a hole cut out of the paper.

So I went back to the major who was supposed to be in charge of all this and complained. You know, this took a little time, but I felt I was sort of the representative to get the thing straightened out. The major tried to explain to me that these people who were the censors had been taught how to do it, but they didn't understand this new way that we had to be so delicate about.

So, anyway, he said, "What's the matter, don't you think I have good will?"

I said, "Yes, you have perfectly good will but I don't think you have *power*." Because, you see, he had already been on the job three or four days.

He said, "We'll see about *that*!" He grabs the telephone, and everything is straightened out. No more is the letter cut.

However, there were a number of other difficulties. For example, one day I got a letter from my wife and a note from the censor that said, "There was a code enclosed without the key, and so we removed it."

So when I went to see my wife in Albuquerque that day, she said, "Well, where's all the stuff?"

I said, "What stuff?"

She said, "Litharge, glycerine, hot dogs, laundry."

I said, "Wait a minute – that was a list?"

She said, "Yes."

"That was a *code*," I said. "They thought it was a code – litharge, glycerine, etc." (She wanted litharge and glycerine to make a cement to fix an onyx box.)

All this went on in the first weeks before we got each other straightened out. Anyway, one day I'm piddling around with the computing machine, and I notice something very peculiar. If you take 1 divided by 243 you get .004115226337 ... It's quite cute, and then it goes a little cockeyed when you're carrying; confusion occurs for only about three numbers, and then you can see how the 10 10 13 is really equivalent to 114 again, or 115 again, and it keeps on going, and repeats itself nicely after a couple of cycles. I thought it was kind of amusing.

Well, I put that in the mail, and it comes back to me. It doesn't go through, and there's little note: "Look at Paragraph 17B." I look Paragraph 17B. It says, "Letters are to be written only in English, Russian, Spanish, Portuguese, Latin, German, and so forth. Permission to use any other language must be obtained in writing." And then it said, "No codes."

So I wrote back to the censor a little note included in my letter which said that I feel that of course this cannot be a code, because if you actually *do* divide 1 by 243 you do, in fact, *get* all that, and therefore there's no more information in the number .004115226337 . . . than there is in the number 243 — which is hardly any information at all. And so forth. I therefore asked for permission to use Arabic numerals in my letters. So, I got that through all right.

There was always some kind of difficulty with the letters going back and forth. For example, my wife kept mentioning the fact that she felt uncomfortable writing with the feeling that the censor was looking over her shoulder. Now, as a rule, we weren't supposed to mention censorship. *We* weren't , but how could they tell *her*? So they kept sending me a note: "Your wife mentioned censorship." *Certainly* my wife mentioned censorship. So finally they sent me a note that said, "Please inform your wife not to mention censorship in her letters." So I start my letter: "I have been instructed to inform you not to mention censorship in your letters." *Phoom, Phoooom*, it comes right back! So I write, "I have been instructed to inform my wife not to mention censorship. How in the heck am I going to do it? Furthermore, *why* do I have to instruct her not to mention censorship? You keeping something from me?"

It is very interesting that the censor himself has to tell me to tell my wife not to tell me that she's. . . . But they had an answer. They said, yes, that they are worried about mail being intercepted on the way from Albuquerque, and that someone might find out that there was censorship if they looked in the mail, and would she please act much more normal.

So I went down the next time to Albuquerque, and I talked to her and I said, "Now, look, let's not mention censorship." But we had had so much trouble that we at last worked out a code, something illegal. If I would put a dot at the end of my signature, it meant I had had trouble again, and she would move on to the next of the moves that she had concocted. She would sit there all day long, because she was ill, and she would think of things to do. The last thing she did was to send me an advertisement which she found perfectly legitimately. It said, "Send your boyfriend a letter on a jigsaw puzzle. We sell you the blank, you write the letter on it, take it all apart, put it in a little sack, and mail it." I received that one with a note saying, "We do not have time to play games. Please instruct your wife to confine herself to ordinary letters."

Well, we were ready with the one more dot, but they straightened out just in time and we didn't have to use it. The thing we had ready for the next one was that the letter would start, "I hope you remembered to open this letter carefully because I have included the Pepto Bismol powder for your stomach as we arranged." It would be a letter full of powder. In the office we expected they would open it quickly, the powder would go all over the floor, and they would get all upset because you are not supposed to upset anything. They'd have to gather up all this Pepto Bismol ... But we didn't have to use that one. OK?

As a result of all these experiences with the censor, I knew exactly what could get through and what could not get through. Nobody else knew as well as I. And so I made a little money out of all of this by making bets.

One day I discovered that the workmen who lived further out and wanted to come in were too lazy to go around through the gate, and so they had cut themselves a hole in the fence. So I went out the gate, went over to the hole and came in, went out again, and so on, until the sergeant at the gate begins to wonder what's happening. How come this guy is always going out and never coming in? And, of course, his natural reaction was to call the lieutenant and try to put me in jail for doing this. I explained that there was a hole.

You see, I was always trying to straighten people out. And so I made a bet with somebody that I could tell about the hole in the fence in a letter, and mail it out. And sure enough, I did. And the way I did it was I said, "You should see the way they administer this place (that's what we were *allowed* to say). There's a hole in the fence 71 feet away from such and such a place, that's this size and that size, that you can walk through."

Now, what can they do? They can't say to me that there is no such hole? I mean, what are they going to do? It's their own hard luck that there's such a hole. They should *fix* the hole. So I got that one through.

I also got through a letter that told about how one of the boys who worked in one of my groups, John Kemeny, had been wakened up in the middle of the night and grilled with lights in front of him by some idiots in the Army there because they found out something about his father, who was supposed to be a communist or something. Kemeny is a famous man now.

Well, there were other things. Like the hole in the fence, I was always trying to point these things out in a non-direct manner. And one of the things I wanted to point out was this — that at the very beginning we had terribly important secrets; we'd worked out lots of stuff about bombs and uranium and how it worked, and so on; and all this stuff was in documents that were in wooden filing cabinets that had little, ordinary, common padlocks on them. Of course, there were various things made by the shop — like a rod that would go down and then a padlock to hold it, but it was always just a padlock. Furthermore, you could get the stuff out without even opening the padlock. You just tilt the cabinet over backwards. The bottom drawer has a little rod that's supposed to hold the papers together, and there's a long wide hole in the wood underneath. You can pull the papers out from below.

So I used to pick the locks all the time and point out that it was very easy to do. And every time we had a meeting of everybody together, I would get up and say that we have important secrets and we shouldn't keep them in such things; we need better locks. One day Teller got up at the meeting, and he said to me, "Well, I don't keep my most important secrets in my filing cabinet; I keep them in my desk drawer. Isn't that better?"

I said, "I don't know. I haven't seen your desk drawer."

Well, he was sitting near the front of the meeting, and I'm sitting further back. So the meeting continues, and I sneak out and go down to see his desk drawer. OK?

I don't even have to pick the lock on the desk drawer. It turns out that if you put your hand in the back, underneath, you can pull out the paper like those toilet paper dispensers. You pull out one, it pulls another, it pulls another . . . I emptied the whole damn drawer, put everything away to one side, and went back upstairs.

The meeting was just ending, and everybody was coming out, and I joined the crew and ran to catch up with Teller, and I said, "Oh, by the way, let me see your desk drawer."

"Certainly," he said, and he showed me the desk.

I looked at it and said, "That looks pretty good to me. Let's see what you have in there."

"I'll be very glad to show it to you," he said, putting in the key and opening the drawer. "If," he said, "you hadn't already seen it yourself."

The trouble with playing a trick on a highly intelligent man like Mr. Teller is that the *time* it takes him to figure out from the moment that he sees there is something wrong till he understands exactly what happened is too damn small to give you any pleasure!

After I was able to open the filing cabinets by picking the locks, they got filing cabinets that had safe combinations. Now, one of my diseases, one of my things in life, is that anything that is secret I try to undo. And so the locks to those filing cabinets represented a challenge to me. How the hell to open them? So I worked and worked on them. There are all kinds of stories about how you can feel the numbers and listen to things and so on. That's true; I understand it very well — for old fashioned safes. But these had a new design so that nothing would be pushing against the wheels while you were trying them, and none of the old methods would work.

I read books by locksmiths, which always say in the beginning how they opened the locks when the safe is under water and the woman in it is drowning or something, and the great locksmith opened the safe. And then in the back they tell you how they do it, and they don't tell you anything sensible. It doesn't sound like they could really open safes that way — like *guess* the combination on the basis of the psychology of the person who owns it! So I always figured they were keeping the method a secret, and like a kind of disease, I kept working on these things until I found out a few things.

First, I found out how big a range you need to open the combination, how close you have to be. And then I invented a system by which you could try all the necessary combinations — 8,000, as it turned out, because you could be within two of every number. And then I worked out a scheme by which I could try numbers without altering a number that I had already set, by correctly moving the wheels, so that I could try all the combinations in eight hours. And then finally I discovered (this took me about two years of researching) that it's easy to take the last two numbers of the combination off the safe if the safe is open. If the drawer was pulled out, you could turn the number and see the bolt go up and play around and find out what number it comes back at, and stuff like that. With a little trickery, you can get the combination off.

So I used to practice it like a cardsharp practices cards, you know — all the time. Quicker and quicker and more and more unobtrusively I would come in and talk to some guy. I'd sort of lean against his filing cabinet, and you wouldn't even notice I'm doing anything. I'm not doing anything — just playing with the dial, that's all, just playing with the dial. But all the time I was taking the two numbers off! And then I would go back to my office and write the two numbers down, the last two numbers of the three. Now, if you have the last two numbers, it takes just a minute to try for the first number; there's only 20 possibilities, and it's open. OK? It takes about three minutes to open a safe if you know the last two numbers.

So I got an excellent reputation for safe-cracking. They would say to me, "Mr. Schmultz is out of town, and we need a document from his safe. Can you open it?"

I'd say, "Yes, I can open it, but I have to go get my tools."

I didn't need any tools, but I'd go to my office and look up the number of his safe. I had the last two numbers for everybody's safe in my office. I'd put a screwdriver in my back pocket to account for the tool I claimed I needed.

I'd go back to the room and close the door. The attitude is that this business about how you open safes is not something that everybody should know because it makes everything very unsafe. So I'd close the door and then sit down and read a magazine or do something. I'd average about 20 minutes of doing nothing, and then I'd open it. Well, I really opened it right away to see that everything was all right, and then I'd sit there for 20 minutes to give myself a good reputation that it wasn't too easy, that there was no trick to it. And then I'd come out, sweating a bit, and say, "It's open. There you are."

Once, however, I did open a safe purely by accident, and that helped to reinforce my reputation. It was a sensation, but it was pure luck.

I went back to Los Alamos after the war was over to finish some papers, and there I did some safe opening that — well, I could write a safecracker book *better* than any previous safecracker book. It would start by explaining how I opened the safe — absolutely cold, without knowing the combination — which contained *more* secret things than any safe that's ever been opened. I opened the safe that contained the secret of the atomic bomb — *all* the secrets, the formulas, the rates at which neutrons are liberated from uranium, how much uranium you need to make a bomb, how much was being made and available, all the theories, all the calculations, the WHOLE DAMN THING!

This is the way it was done.

I was trying to write a report. I needed some material but it was a Saturday. I thought everybody worked. I thought it was like Los Alamos *used* to be. So I went down to get some documents from the library. The library at Los Alamos had all these documents in a great vault with a lock and dial of a kind I didn't know anything about. Filing cabinets I understood, but I was an expert only on filing cabinets. Not only that, but there were guards walking back and forth in front with guns. I couldn't get that vault open. OK?

But then I thought, wait! Old Freddy DeHoffman is in charge of deciding which documents now can be de-classified. He had to run down to the library and back so often, he got tired of it. And he got a brilliant idea. He would get a copy made of every document in the Los Alamos library. And he'd stick them in *his* files. He had *nine* filing cabinets, one right next to the other in two rooms, full of all the documents of Los Alamos.

I went up to his office. The office door was open. It looked like he was coming back any minute; the light was lit. So I waited. And, as always when I'm waiting, I diddled the knobs. I tried 10-20-30 — didn't work. I tried 20-40-60 — didn't work. I tried everything, because I'm waiting, with nothing to do.

Then I began to think. You know, I have never been able to figure out how to open safes cleverly, so maybe those locksmith people don't either.

Maybe all the stuff they tell me about psychology is right. I'm going to open this one by psychology.

The first thing the book says is: "The secretary is very often nervous that she will forget the combination." She's been told the combination, but she might forget, and the boss might forget. She has to know. So she nervously writes it somewhere. Where? List of places where a secretary might write combinations, OK? It starts right out with the most clever thing: You open the drawer, and on the wood along the outside of the drawer is written carelessly a number, as if it is an invoice number. That's the combination number. So. It's on the side of the desk, OK? I remembered that; it's in the book.

The desk drawer was locked, but I picked the lock easily. I pulled out the drawer, looked along the wood. Nothing. All right. There were a lot of papers in the drawer. I fished around among the papers, and finally I found it, a nice little piece of paper which has the Greek alphabet — alpha, beta, gamma, delta, and so forth — carefully printed.

The secretaries have to know how to make those letters and what to call them when they're talking about them, right? So they each had a copy of the thing. But — carelessly scrawled across the top is, *pi is equal to 3.14159*. Well, why does she needed the numerical value of pi? She's not computing anything. So I walked up to the safe. 31-41-59 — doesn't open. 13-14-95 — doesn't open. 95-14-13 — doesn't open. For 20 minutes I turned pi upside down. Nothing happened.

So I started walking out of the office, and I remembered in the book about the psychology, and I said, "You know, it's true. Psychologically, DeHoffman is *just* the kind of a guy to use a mathematical constant for his safe combination. And the other important mathematical constant is *e*." So I walk back to the safe. 27-18-28 — click, clock, it opens.

I checked, by the way, that all the rest of the filing cabinets had the same combination.

Well, I want to tell about some of the special problems I had at Los Alamos that were rather interesting. One thing had to do with the safety of the plant at Oak Ridge. Los Alamos was going to make the bomb, but at Oak Ridge they were trying to separate the isotopes of uranium — uranium 238 and uranium 235, the explosive one. They were *just* beginning to get infinitesimal amounts from an experimental thing of 235, and the same time they were practicing the chemistry. There was going to be a big plant, they were going to have vats of the stuff, and then they were going to take the purified stuff and repurify and get it ready for the next stage. (You have to purify it in several stages.) So they were practicing on the one hand, and they were just getting a little bit of U235 from one of the pieces of apparatus experimentally on the other hand. And they were trying to learn how to assay it, to determine how much uranium 235 there is in it — and though we would send them instructions, they never got it right.

So finally Segrè said that the only possible way to get it right was for him to go down there and see what they were doing. The Army people said, "No, it is our policy to keep all the information of Los Alamos at one place."

The people in Oak Ridge didn't know anything about what it was to be used for; they just knew what they were trying to do. I mean the higher people knew they were separating uranium, but they didn't know how powerful the bomb was, or exactly how it worked or anything. The people underneath didn't know at *all* what they were doing. And the Army wanted to keep it that way. There was no information going back and forth. But Segrè insisted they'd never get the assays right, and the whole thing would go up in smoke. So he finally went down to see what they were doing, and as he was walking through he saw them wheeling a tank carboy of water, green water — which is uranium nitrate solution.

He says, "Uh, you're going to handle it like that when it's purified too? Is that what you're going to do?"

They said, "Sure — why not?"

"Won't it explode?" he says.

Huh! *Explode?*

And so the Army said, "You see! We shouldn't have let any information get to them! Now they are all upset."

Well, it turned out that the Army had realized how much stuff we needed to make a bomb — 20 kilograms or whatever it was — and they realized that this much material, purified, would never be in the plant, so there was no danger. But they did *not* know that the neutrons were enormously more effective when they are slowed down in water. And so in water it takes less than a tenth — no, a hundredth — as much material to make a reaction that makes radioactivity. It kills people around and so on. So, it was *very* dangerous, and they had not paid any attention to the safety at all.

So a telegram goes from Oppenheimer to Segrè: "Go through the entire plant. Notice where all the concentrations are supposed to be, with the process as *they* designed it. We will calculate in the meantime how much material can come together before there's an explosion."

Two groups started working on it. Christy's group worked on water solutions and my group worked on dry powder in boxes. We calculated about how much material they could accumulate safely. And Christy was going to go down and tell them all at Oak Ridge what the situation was, because this whole thing is broken down and we *have* to go down and tell them now. So I happily gave all my numbers to Christy, and said, "You have all the stuff, so go." Christy got pneumonia; I had to go.

I never traveled on an airplane before. I traveled on an airplane. They strapped the secrets in a little thing on my back! The airplane in those days was like a bus, except the stations were further apart. You stopped off every once in a while to wait.

There was a guy standing there next to me swinging a chain, saying something like, "It must be *terribly* difficult to fly without a priority on airplanes these days."

I couldn't resist. I said, "Well, I don't know. I *have* a priority."

A little bit later he tried again. "It looks like this. There are some generals coming. They are going to put off some of us number 3's."

"It's all right," I said, "I'm a number 2."

He probably wrote to his congressman — if he wasn't a congressman himself — saying, "What are they doing sending these little kids around with number 2 priorities in the middle of the war?"

At any rate, I arrived at Oak Ridge. The first thing I did was have them take me to the plant, and I said nothing, I just looked at everything. I found out that the situation was even worse than Segrè reported because he noticed certain boxes in big lots in a room, but he didn't notice a lot of boxes in another room on the other side of the same wall — and things like that. Now, if you have too much stuff together, it goes up, you see.

So I went through the entire plant. I have a very bad memory, but when I work intensively I have a good short-term memory, and so I could remember all kinds of crazy things like building 90-207, vat number so and so, and so forth.

I went home that night, and I went through the whole thing, explained where all the dangers were, and what you would have to do to fix this. It's rather easy. You put cadmium in solutions to absorb the neutrons in the water, and you separate the boxes so they are not too dense, according to certain rules.

The next day there was going to be a big meeting. I forgot to say that before I left Los Alamos Oppenheimer said to me, "Now, the following people are technically able down there at Oak Ridge: Mr. Julian Webb, Mr. so and so, and so on. I want you to make sure that these people are at the meeting, that you tell them how the thing can be made safe, so that they really *understand*."

I said, "What if they're not at the meeting? What am I supposed to do?"

He said, "Then you should say: *Los Alamos cannot accept the responsibility for the safety of the Oak Ridge plant* unless _____!"

I said, "You mean me, little Richard, is going to go in there and say —?"

He said, "Yes, little Richard, you go and do that."

I really grew up fast!

So when I arrived, sure enough, the big shots in the company and the technical people that I wanted were there, and the generals and everyone who was interested in this very serious problem. And that was good because the plant would have blown up if nobody had paid attention to this problem.

Well, there was a Lieutenant Zumwalt who took care of me, and he told me that the colonel said I shouldn't tell them how the neutrons work and all the details because we want to keep things separate, so just tell them what to do to keep it safe.

I said, "In my opinion it is impossible for them to obey a bunch of rules unless they understand how it works. So it's my opinion that it's only going to work if I tell them, and *Los Alamos cannot accept the responsibility for the safety of the Oak Ridge plant unless they are fully informed as to how it works!*"

It was great. The lieutenant takes me to the colonel and repeats my remark. The colonel says, "Just five minutes," and then he goes to the window and he stops and thinks. That's what they're very good at — making decisions. I thought it was very remarkable how a problem of whether or not information as to how the bomb works should be in the Oak Ridge plant or not had to be decided and *could* be decided in five minutes. So I have a great deal of respect for these military guys, because I never can decide anything very important in any length of time of all.

` So in five minutes he said, "All right, Mr. Feynman, go ahead."

So I sat down and I told them all about neutrons, how they worked, da da, ta ta ta, there are too many neutrons together, you've got to keep the material apart, cadmium absorbs, and slow neutrons are more effective than fast neutrons, and yak yak — all of which was elementary stuff at Los Alamos, but they had never heard of any of it, so I turned out to be a tremendous genius to them.

I was a god coming down from the sky! Here were all these phenomena that were not understood and never heard of before — but I knew all about it; I could give them facts and numbers and everything else. So, from being rather primitive back there at Los Alamos, I became a super-genius at the other end.

The result was that they decided to set up little groups to make their own calculations to learn how to do it. They started to re-design plants, and the designers of the plants were there, the construction designers, and engineers, and chemical engineers for the new plant that was going to handle the separated material.

They told me to come back in a few months, so I came back when the engineers had finished the design of the plant. Now it was for me to look at the plant. OK?

How do you look at a plant that ain't built yet? I don't know. Well, Lieutenant Zumwalt, who was always coming around with me because I had to have an escort everywhere, takes me into this room where there are these two engineers and a *looooong* table covered with a stack of large, long blueprints representing the various floors of the proposed plant.

I took mechanical drawing when I was in school, but I am not good at reading blueprints. So they start to explain it to me, because they think I am a genius. Now, one of the things they had to avoid in the plant was accumulation. So they had problems like when there's an evaporator working, which is trying to accumulate the stuff, if the valve gets stuck or something like that and too much stuff accumulates, it'll explode. So they explained to me that this plant is designed so that if any one valve gets stuck nothing will happen. It needs at least two valves everywhere.

Then they explain how it works. The carbon tetrachloride comes in here, the uranium nitrate from here comes in here, it goes up and down, it goes up through the floor, comes up through the pipes, coming up from the second floor, *bluuuuurp* – going through the stack of blueprints, down-up-down-up, talking very fast, explaining the very, very complicated chemical plant.

I'm completely dazed. Worse, I don't know what the symbols on the blueprint mean! There is some kind of a thing that at first I think is a window. It's a square with a little cross in the middle, all over the damn place. I think it's a window, but no, it can't be a window, because it isn't always at the edge. I want to ask them what it is.

You must have been in a situation like this when you didn't ask them right away. Right away it would have been OK. But now they've been talking a little bit too long. You hesitated too long. If you ask them now they'll say, "What are you wasting my time all this time for?"

I don't know what to do. (You are not going to believe this story, but I swear it's absolutely true – it's such sensational luck.) I thought, what am I going to *do?* I got an idea. Maybe it's a valve? So, in order to find out whether it's a valve or not, I take my finger and I put it down on one of the mysterious little crosses in the middle of one of the blueprints on page number 3, and I say, "What happens if this valve gets stuck?" — figuring they're going to say, "That's not a valve, sir, that's a window."

So one looks at the other and says, "Well, if *that* valve gets stuck —" and he goes up and down on the blueprint, up and down, the other guy up and down, back and forth, back and forth, and they both look at each other and they *tchk, tchk, tchk,* and they turn around to me and they open their mouths like astonished fish and say, "You're absolutely right, sir."

So they rolled up the blueprints and away they went and we walked out. And Mr. Zumwalt, who had been following me all the way through, said, "You're a genius. I got the idea you were a genius when you went through the plant once and you could tell them about evaporator C-21 in building 90-207 the next morning," he says, "but what you have just done is so *fantastic* I want to know how, *how* do you do that?"

I told him you try to find out whether it's a valve or not.

Well, another kind of problem I worked on was this. We had to do lots of calculations, and we did them on Marchant calculating machines. By the way, just to give you an idea of what Los Alamos was like: We had these Marchant computers — hand calculators with numbers. You push them, and they multiply, divide, add and so on, but not easy like they do now. They were mechanical gadgets, failing often, and they had to be sent back to the factory to be repaired. Pretty soon you were running out of machines. So a few of us started to take the covers off. (We weren't supposed to. The rules read: "You take the covers off, we cannot be responsible. . .") So we took the covers off and we got a nice series of lessons on how to fix them, and we got better and better at it as we got more and more elaborate repairs. When we got something too complicated, we sent it back to the factory, but we'd do the easy ones and kept the things going. I ended up doing all the computers and there was a guy in the machine shop who took care of typewriters.

Anyway, we decided that the big problem — which was to figure out exactly what happened during the bomb's explosion, so you can figure out exactly how much energy was released and so on — required much more calculating than we were capable of. A rather clever fellow by the name of Stanley Frankel realized that it could possibly be done on IBM machines. The IBM company had machines for business purposes, adding machines called tabulators for listing sums, and a multiplier that you put cards in and it would take two numbers from a card and multiply them. There were also collators and sorters and so on.

So Frankel figured out a nice program. If we got enough of these machines in a room, we could take the cards and put them through a cycle. Everybody who does numerical calculations now knows exactly what I'm talking about, but this was kind of a new thing then — mass production with machines. We had done things like this on adding machines. Usually you go one step across, doing everything yourself. But this was different — where you go first to the adder, then to the multiplier, then to the adder, and so on. So Frankel designed this system and ordered the machines from the IBM company, because we realized it was a good way of solving our problems.

We needed a man to repair the machines, to keep them going and everything. And the Army was always going to send this fellow they had, but he was always delayed. Now, we *always* were in a hurry. *Everything* we did, we tried to do as quickly as possible. In this particular case, we worked out all the numerical steps that the machines were supposed to do — multiply this, and then do this, and subtract that. Then we worked out the program, but we didn't have any machine to test it on. So we set up this room with girls in it. Each one had a Marchant. But *she* was the multiplier, and *she* was the adder, and this one cubed, and we had index cards, and all she did was cube this number and send it to the next one.

We went through our cycle this way until we got all the bugs out. Well, it turned out that the speed at which we were able to do it was a hell of a lot faster than the other way, where every single person did all the steps. We got speed with this system that was the predicted speed for the IBM machine. The only difference is that the IBM machines didn't get tired and could work three shifts. But the girls got tired after a while.

Anyway, we got the bugs out during this process, and finally the machines arrived, but not the repairman. These were some of the most complicated machines of the technology of those days, big things that came partially disassembled, with lots of wires and blueprints of what to do. We went down and we put them together, Stan Frankel and I and another fellow, and we had our troubles. Most of the trouble was the big shots coming in all the time and saying, "You're going to break something!"

We put them together, and sometimes they would work, and sometimes they were put together wrong and they didn't work. Finally I was working on some multiplier and I saw a bent part inside, but I was afraid to straighten it because it might snap off — and they were always telling us we were going to bust something irreversibly. When the repairman finally got there, he fixed the machines we hadn't got ready, and everything was going. But he had trouble with the one that I had had trouble with. So after three days he was still working on that *one* last machine.

I went down, I said, "Oh, I noticed that was bent."

He said, "Oh, of course. That's all there is to it!" *Bend!* It was all right. So that was it.

Well, Mr. Frankel, who started this program, began to suffer from the computer disease that anybody who works with computers now knows about. It's a very serious disease and it interferes completely with the work. The trouble with computers is you *play* with them. They are so wonderful. You have these switches — if it's an even number you do this, if it's an odd number you do that — and pretty soon you can do more and more elaborate things if you are clever enough, on one machine.

And so after a while the whole system broke down. Frankel wasn't paying any attention; he wasn't supervising anybody. The system was going very, very slowly — while he was sitting in a room figuring out how to make one tabulator automatically print arc-tangent X, and then it would start and it would print columns and then *bitsi, bitsi, bitsi,* and calculate the arc-tangent automatically by integrating as it went along and make a whole table in one operation.

Absolutely useless. We *had* tables of arc-tangents. But if you've ever worked with computers, you understand the disease — the *delight* in being able to see how much you can do. But he got the disease for the first time, the poor fellow who invented the thing.

And so I was asked to stop working on the stuff I was doing in my group and go down and take over the IBM group, and I tried to avoid the disease. And, although they had done only three problems in nine months, I had a very good group.

The real trouble was that no one had ever told these fellows anything. The Army had selected them from all over the country for a thing called Special Engineer Detachment — clever boys from high school who had engineering ability. They sent them up to Los Alamos. They put them in barracks. And they would tell them *nothing*.

Then they came to work, and what they had to do was work on IBM machines — punching holes, numbers that they didn't understand. Nobody told them what it was. The thing was going very slowly. I said that the first thing there has to be is that these technical guys know what we're doing. Oppenheimer went and talked to the security and got special permission so I could give a nice lecture about what we were doing, and they were all excited: "We're fighting a war! We see what it is!" They knew what the numbers meant. If the pressure came out higher, that meant there was more energy released, and so on and so on. They knew what they were doing.

Complete transformation! *They* began to invent ways of doing it better. They improved the scheme. They worked at night. They didn't need supervising in the night; they didn't need anything. They understood everything; they invented several of the programs that we used — and so forth.

So my boys really came through, and all that had to be done was to tell them what it was, that's all. As a result, although it took them nine months to do three problems before, we did nine problems in *three* months, which is nearly ten times as fast.

But one of the secret ways we did our problems was this: The problems consisted of a bunch of cards that had to go through a cycle. First add, then multiply — and so it went through the cycle of machines in this room, slowly, as it went around and around. So we figured a way to put a different colored set of cards through a cycle too, but out of phase. We'd do two or three problems at a time.

But this got us into *another* problem. Near the end of the war for instance, just before we had to make a test near Alamogordo, the question was: How much energy would be released? We had been calculating the release from various designs, but we hadn't computed for the specific design that was ultimately used. So Bob Christy came down and said, "We would like the results for how this thing is going to work in one month" — or some very short time, like three weeks.

I said, "It's impossible."

He said, "Look, you're putting out nearly two problems a month. It takes only two weeks per problem, or three weeks per problem."

I said, "I know. It really takes much longer to do the problem, but we're doing them in *parallel*. As they go through, it takes a long time and there's no way to make it go around faster."

So he went out, and I began to think. Is there a way to make it go around faster? What if we did nothing else on the machine, so there was nothing else interfering? I put a challenge to the boys on the blackboard — CAN WE DO IT? They all start yelling, "Yes, we'll work double shifts, we'll work overtime," — all this kind of thing. "We'll *try* it. We'll *try* it!"

And so the rule was: All other problems *out*. Only one problem and just concentrate on this one. So they started to work.

My wife died in Albuquerque, and I had to go down. I borrowed Fuchs' car. He was a friend of mine in the dormitory. He had an automobile. He was using the automobile to take the secrets away, you know, down to Santa Fe. He was the spy. I didn't know that. I borrowed his car to go to Albuquerque. The damn thing got three flat tires on the way. I came back from there, and

I went into the room, because I was supposed to be supervising everything, but I couldn't do it for three days.

It was in this *mess*. There's white cards, there's blue cards, there's yellow cards, and I start to say, "You're not supposed to do more than one problem – only one problem!" They said, "Get out, get out, get out. Wait – and we'll explain everything."

So I waited, and what happened was this. As the cards went through, sometimes the machine made a mistake, or they put a wrong number in. What we used to have to do when that happened was to go back and do it over again. But they noticed that a mistake made at some point in one cycle only affects the nearby numbers, the next cycle affects the nearby numbers, and so on. It works its way through the pack of cards. If you have 50 cards and you make a mistake at card number 39, it affects 37, 38, and 39. The next, card 36, 37, 38, 39, and 40. The next time it spreads like a disease.

So they found an error back a way, and they got an idea. They would only compute a small deck of 10 cards around the error. And because 10 cards could be put through the machine faster than the deck of 50 cards, they would go rapidly through with this other deck while they continued with the 50 cards with the disease spreading. But the other thing was computing faster, and they would seal it all up and correct it. OK? Very clever.

That was the way those guys worked, really hard, very clever, to get speed. There was no other way. If they had to stop to try to fix it, we'd have lost time. We couldn't have got it. That was what they were doing.

Of course, you know what happened while they were doing that. They found an error in the blue deck. And so they had a yellow deck with a little fewer cards; it was going around faster than the blue deck. Just when they are going crazy – because after they get this straightened out, they have to fix the white deck – the *boss* comes walking in.

"Leave us alone," they say. So I left them alone and everything came out. We solved the problem in time and that's the way it was.

I would like to tell a little about some of the people I met. I was an underling at the beginning. I became a group leader. But I met some very great men. It is one of the great experiences of my life to have met all these wonderful physicists.

There was, of course, Fermi. He came down once from Chicago, to consult a little bit, to help us if we had some problems. We had a meeting with him, and I had been doing some calculations and gotten some results. The calculations were so elaborate it was very difficult. Now, usually I was the expert at this; I could always tell you what the answer was going to look like, or when I got it I could explain why. But this thing was so complicated I couldn't explain *why* it was like that.

So I told Fermi I was doing this problem, and I started to describe the results. He said, "Wait, before you tell me the result, let me think. It's going to come out like this (he was right), and it's going to come out like this because of so and so. And there's a perfectly obvious explanation for this —."

He was doing what I was supposed to be good at, ten times better. So that was quite a lesson to me.

Then there was Von Neumann, the great mathematician. We used to go for walks on Sunday. We'd walk in the canyons, and we'd often walk with Bethe, and Von Neumann, and Bacher. It was a great pleasure. And Von Neumann gave me an interesting idea; that you don't have to be responsible for the world that you're in. So I have developed a very powerful sense of social irresponsibility as a result of Von Neumann's advice. It's made me a very happy man ever since. But it was Von Neumann who put the seed in that grew into my *active* irresponsibility!

I also met Niels Bohr. His name was Nicholas Baker in those days, and he came to Los Alamos with Jim Baker, his son, whose name is really Aage Bohr. They came from Denmark, and they were *very* famous physicists, as you know. Even to the big shot guys, Bohr was a great god.

We were at a meeting once, the first time he came, and everybody wanted to *see* the great Bohr. So there were a lot of people there, and we were discussing the problems of the bomb. I was back in a corner somewhere. He came and went, and all I could see of him was from between people's heads, from the corner.

In the morning of the day he's due to come next time, I get a telephone call.

"Hello — Feynman?"

"Yes."

"This is Jim Baker." It's his son. "My father and I would like to speak to you."

"Me? I'm Feynman, I'm just a —."

"That's right. OK."

So, at 8 o'clock in the morning, before anybody's awake, I go down to the place. We go into an office in the technical area and he says, "We have been thinking how we could make the bomb more efficient and we think of the following idea."

I say, "No, it's not going to work. It's not efficient. Blah, blah, blah."

So he says, "How about so and so?"

I said, "That sounds a little bit better, but it's got this damn fool idea in it."

So forth, back and forth. I was always *dumb* about one thing. I never knew who I was talking to. I was always worried about the physics. If the idea looked lousy, I said it looked lousy. If it looked good, I said it looked good. Simple proposition.

I've always lived that way. It's nice, it's pleasant — if you can do it. I'm lucky. Just as I was lucky with that blueprint, I'm lucky in my life that I can do this.

So, this went on for about two hours, going back and forth over lots of ideas, back and forth, arguing. The great Niels kept lighting his pipe; it always went out. And he talked in a way that was un-understandable — mumble, mumble, hard to understand. His son I could understand better.

"Well," he says finally, lighting his pipe, "I guess we can call in the big shots *now*." So then they called all the other guys and had a discussion with them.

Then the son told me what happened. The last time he was there, he said to his son, "Remember the name of that little fellow in the back over there? He's the only guy who's not afraid of me, and will say when I've got a crazy idea. So *next* time when we want to discuss ideas, we're not going to be able to do it with these guys who say everything is yes, yes, Dr. Bohr. Get that guy and we'll talk with him first."

The next thing that happened, of course, was the test, after we'd made the calculations. I was actually at home on a short vacation at that time, after my wife died, and so I got a message that said, "The baby is expected on such and such a day."

I flew back, and I *just* arrived when the buses were leaving, so I went straight out to the site and we waited out there, 20 miles away. We had a radio, and they were supposed to tell us when the thing was going to go off and so forth, but the radio wouldn't work, so we never knew what was happening. But just a few minutes before it was supposed to go off the radio started to work, and they told us there was 20 seconds or something to go.

For people who were far away like we were — others were closer, 6 miles away — they gave out dark glasses that you could watch it with. Dark glasses! Twenty miles away, you couldn't see a damn thing through dark glasses. So I figured the only thing that could really hurt your eyes — bright light can never hurt your eyes — is ultraviolet light. I got behind a truck windshield, because the ultraviolet can't go through glass, so that would be safe, and so I could *see* the damn thing. OK.

Time comes, and this *tremendous* flash out there is so bright that I duck, and I see this purple splotch on the floor of the truck. I said, "That ain't it. That's an after-image." So I look back up, and I see this white light changing into yellow and then into orange. The clouds form and then they disappear again; the compression and the expansion forms and makes clouds disappear. Then finally a big ball of orange, the center that was so bright, becomes a ball of orange that starts to rise and billow a little bit and get a little black around the edges, and then you see it's a big ball of smoke with flashes on the inside of the fire going out, the heat.

All this took about one minute. It was a series from bright to dark, and I had *seen* it. I am about the only guy who actually looked at the damn thing — the first Trinity test. Everybody else had dark glasses, and the people at six miles couldn't see it because they were all told to lie on the floor. I'm probably the only guy who saw it with the human eye.

Finally, after about a minute and a half, there's suddenly a tremendous noise — *BANG*, and then a rumble, like thunder — and that's what convinced me. Nobody had said a word during this whole thing. We were all just watching quietly. But this sound released everybody — released me particularly because the solidity of the sound at that distance meant that it had really worked.

The man standing next to me said, "What's that?"

I said, "That was the bomb."

The man was William Laurence of the *New York Times*. He was there to write an article describing the whole situation. I had been the one who was supposed to have taken him around. Then it was found that it was too technical for him, and so later Mr. Smyth of Princeton came and I showed him around. One thing we did, we went into a room and there on the end of a narrow pedestal was a small silver-plated ball. You could put your hand on it. It was warm. It was radioactive. It was plutonium. And we stood at the door of this room, talking about it. This was a new element that was made by man, that had never existed on the earth before, except for a very short period possibly at the very beginning. And here it was all isolated and radioactive and had these properties. And we had made it. And so it was *tremendously* valuable.

Meanwhile, you know how people do when they talk — you kind of jiggle around and so forth. He's kicking the doorstop, you see, and I said, "Yes, the doorstop certainly is appropriate for this door." The doorstop was a hemisphere of yellowish metal — gold, as a matter of fact.

What had happened was that we needed to do an experiment to see how many neutrons were reflected by different materials in order to save the neutrons so we didn't use so much fissionable material. We had tested many different materials. We had tested platinum, we had tested zinc, we had tested brass, we had tested gold. So, in making the tests with the gold, we had these pieces of gold and somebody had the clever idea of using that great ball of gold for a doorstop for the door of the room that contained the plutonium.

After the thing went off, there was tremendous excitement at Los Alamos. Everybody had parties, we all ran around. I sat on the end of a jeep and beat drums and so on. But one man I remember, Bob Wilson, was just sitting there moping.

I said, "What are you moping about?"

He said, "It's a terrible thing that we made."

I said, "But you started it. You got us into it."

You see, what happened to me — what happened to the rest of us — is we *started* for a good reason, then you're working very hard to accomplish something and it's a pleasure, it's excitement. And you stop thinking, you know; you just *stop*. So Bob Wilson was the only one who was still thinking about it, at that moment.

I returned to civilization shortly after that and went to Cornell to teach, and my first impression was a very strange one. I can't understand it anymore, but I felt very strongly then. I sat in a restaurant in New York, for example, and I looked out at the buildings and I began to think, you know, about how much the radius of the Hiroshima bomb damage was and so forth . . . How far from here was 34th St? . . . All those buildings, all smashed — and so on. And I would go along and I would see people building a bridge, or they'd be making a new road, and I thought, they're *crazy*, they just don't understand, they don't *understand*. Why are they making new things? It's so useless.

But, fortunately, it's been useless for about 30 years now, isn't it? So I've been wrong for 30 years about it being useless making bridges and I'm glad that those other people had the sense to go ahead.

BERNICE BRODE

TALES OF LOS ALAMOS

I was in Los Alamos from September, 1943, until December 7, 1945. I've been up there twice to visit later, but that's the only time I lived there. My husband, however, was there from early in 1943 until 1946; he went back after he took us home. Let me begin my story by quoting a remark made by Oppenheimer: "The notion of disappearing into the desert for an indeterminate period and under quasi-military auspices disturbed a good many scientists and the families of many more." However true that may have been, it was not the entire picture, and I shall try to present a more balanced description.

In September 1943, the Brode family, including two boys of 11 and 12, set forth in our old Ford to disappear into the desert, to a place not shown on any map. Our first official stop was at 109, East Palace, in Santa Fe, the regional office of the Manhattan District. I was surprised to see a fresh sign over the archway reading, U.S. Army Corps of Engineers, for we had been told that fact in itself was to be unmentionable (a few months later I noticed the sign was taken down). In the street in front of 109, and taking up more than its share of the ancient lane, a dilapidated school bus marked U.S. Army in bright paint was parked. A big soldier was good-naturedly loading his bus with awkward household purchases such as brooms, mops, mirrors, potted plants and kiddie-cars. He was taking orders from pert little housewives with toddlers in tow to "be careful for goodness sake." This was the daily bus, going back to Los Alamos, making two round trips a day, morning and afternoon. We parked behind it and watched the scene. "For crying out loud, ladies," the soldier was saying, "we've got a war on remember? For this I joined the Army!" He threw up his hands in mock despair, and carefully lifted a small, whiney boy into the bus. Then he shouted like a circus barker, "bus leaving for the wilderness up yonder everybody, all aboard, hurry, hurry!" There was another wild scramble with too many children and too many parcels, all pushing into the sagging bus. Then the G.I. fished a length of rope from his pocket, boarded the bus himself and tied the door handle to the front window. As I watched in amazement, he grinned and said, "lady, the General himself told me not to lose any of 'em; seems they're very scarce." The rickety bus rattled off with everyone waving goodby to us as if they were on a picnic.

133

L. Badash, J. O. Hirschfelder and H. P. Broida (eds.), Reminiscences of Los Alamos
1943–1945, 133–159.
Copyright © 1980 by D. Reidel Publishing Company.

Then we went into the courtyard, and through the shabby screen door to the inner office of 109, East Palace. There my husband introduced us to Dorothy McKibbin, who was in charge not only of the office but many other things. She was very lovely, with shining hair and dressed in blue tweed to match her eyes. She had a quiet grace in the midst of all the hubbub. She was a hostess rather than a charge-d'affaires. "Welcome Brode family, and do find chairs while I phone to the site that you have arrived. We have a little peace now that the bus has left. Oh dear, I see Mrs. J. has left her coat; I'll have to remember to send it up on the afternoon bus." She had an air of handling people easily. Only later did I come to know the serious difficulties she had avoided for all of us. The whole place seemed more like a storeroom than an office. There was a minimum of furniture, a few desks and an assortment of old kitchen chairs, with most of the space taken up by stacks of queer looking crates and bundles of local purchases, here, there, and everywhere. I tried not to be too curious, but it was all so different from the red-tape severity of a Washington office; we had just lived in that city the past year. Indeed it was quite nice, just as Oppenheimer had promised. Dorothy made up temporary passes for the boys and me as she chatted, "now make this your headquarters when you come to Santa Fe, everybody does. Leave your parcels here and meet each other. That's what this office is really for." So 109, East Palace, and Dorothy, our only link with Santa Fe, became our private, secret club in the capitol of New Mexico. There we could talk and make plans and have no fear of being overheard. All newcomers passed through this office and were sent up to Los Alamos with their misgivings slightly replaced with trust in the unknown "up yonder." Dorothy was a happy choice for our introduction to the war years on the mesa.

That September day we said goodby to Dorothy and proceeded up the Taos Highway and on to a washboard dirt road to Otowi. My husband began to sing, a sure sign he was confident he was driving his family into something they might enjoy. I was bewitched by the scenery: the stretches of red earth and pink rocks with dark shrubbery scattered along the yellow cliffs, lavender vistas in the distant Sangre de Christo Mountain range. Color was everywhere. Occasionally adobe houses arose from the earth with strings of scarlet chili peppers strung outside to dry. The flat roofs were strewn with ears of yellow, blue, white and dark red corn, also drying for tortillas. We passed the turnoff for San Ildefonso Pueblo with its sacred Black Mesa in the distance, and crossed the narrow road and bridge over the Rio Grande. "Army buses and trucks are not allowed to go over the bridge," said my husband, "they have to go around the longer way, by Española." The boys exclaimed that they could see water through broken planks in the bridge, which explains why the Army replaced it a little bit later. All was beauty and quiet. The rest of the war-torn world had not reached these remote parts. What a superb retreat in which to spend the war years I thought with a rising feeling of guilt, but the thought came too soon. Around the next curve we met our first Army construction work, and a sense of alarm chased away any sense of guilt. Huge bulldozers were tearing down the beautiful rock cliffs and leveling off everything in their paths. Their G.I. drivers seemed unnecessarily happy and purposeful about the destructive work. In the next two years I was to learn that the Army could maintain this bulldozing momentum, conquering and leveling mass after mass, replacing strata of timeless growth with ugly buildings whose purpose was not beauty but grim utility.

Choking, blinking and covered with that layer of dust we were to know so intimately in the next years, we arrived at the east gate. Immediately we were surrounded by serious, armed military police who examined our passes carefully, peeped into our car without a word and then waved us through the gate. The next three miles into the town were noisy and lively with construction and confusion. You could scarcely see due to the dust churned up by vehicles going every direction. "The next stop will be the old stone pumphouse." This was a picturesque remnant of the old boys' school amidst the deafening noise, now, of a rising war factory. Two-story wooden buildings were going up all around the little stone house. On a bit of the old flagstone walk outside we encountered Sally, Mrs. Donald (Moll) Flanders, who had a Dutch haircut and a hardy look about her and talked all in one breath. She welcomed us and ordered Bob to bring us all to the square dance that very evening. Inside the pumphouse, a temporary security office, businesslike WACs took our fingerprints and pressed them on permanent passes which we needed coming and going through the gate. It was soon regarded as a social error to forget one's pass, especially when the others had to wait while someone went back the six-mile round trip to fetch it. Another WAC photographed us in a makeshift booth. We were asked for any identifying scars or oddities of person. These were written on the passes and recorded in the book, rather frightening by implication. Soon we became accustomed to the unusual tests and questions.

Next we went to the housing office, an old wood garage beside the water tower. Mrs. Johnny Williams was in charge and we could hardly greet her in the surrounding madhouse of young people seeking priorities for still-to-be-erected housing units, but Vera was tough and equal to everyone. She laughed and handed us a slip of occupancy of the apartment that Bob had reserved for us. Outside again, we paused to look at the shingled water tower on stilts which marked the center of our town and remains a favorite landmark. Streets had no names and all directions to find houses were given in relation to the water tower. It assumed a symbolic stature in our lives, a reminder of the quiet pre-atomic days before the galloping changes came over the mesa. In winter long icicles hung over the side, to the delight of the small-fry who picked them up. The tower, which still held the water supply for the town, was also to be the crux of our worst community crisis, the water shortage of 1945. Our Ford turned down the road from the water tower in search of our new home, designated T-124C. There, at the far end of the slope where the last of the green Army-constructed clapboard houses were arranged on a pleasantly irregular semi circle, was T-124. Robert Oppenheimer had insisted that the houses follow the natural contours, rather than the straight row formations so dear to the hearts of the Army. General Leslie Groves, head of the Manhattan District, gave in but he grumbled to Oppy, "All this nonsense because the families have to live here. If I could only have my way – put all these scientists in uniform and in barracks. There would be no fuss and feathers." But the General had to put up with the families, and the fuss and feathers.

Three first memories of my impact with T-124, as we alighted in front of the house, remained in my mind. There was a beautiful remote vista of the Jemez Mountains behind the houses; confusion of people, mostly children, in the immediate front area, resembling lower east side New York; and lastly, an unsightly telephone pole right in the middle of an otherwise attractive open space. This proved more unaccountable as time went on for we had no phones in any of the houses at any time. And the pole was very much in the way for parking cars or playing baseball. But then many things were unaccountable in those days. At once our car was besieged by an unofficial welcoming committee from all of the four or five nearby four-family houses. Some were old friends and some unknown neighbors come to have a look at the new residents. We were plunged instantly into the intricacies of mesa life. Mrs. Robert Bacher came running out to tell us of other mutual friends soon to arrive; then she broke off to ask if my boys could babysit for her that very evening! And Alice Smith sounded me out to see if I would accept half of her teaching job at the school which hadn't opened yet. So it went, everyone full of plans and excited enthusiasm. At this point we hadn't even been inside our house. All ages of children were chasing each other upstairs and all over. Barefoot girls were playing hopscotch in the dust, a baseball game was going on in the road, housewives were hanging out clothes on lines strung between houses or attached to trees, a Spanish maid was shaking out a Persian rug from an upper balcony (what did a little extra dust matter?).

It was tall and stately Alice, who lived next door to us, who piloted us inside our kitchen. She had brought over good pans to help furnish our house — it was strictly government issue, plain, functional furniture, that we had until our own furniture arrived months later. We inspected our new abode, going down the narrow hall to the three small bedrooms, wondering how sixteen WACs had ever slept there. There was one tiny bath with funny dull black faucets and a cement and tin-lined shower. Some of the neighbors called the showers the Black Hole of Calcutta. The living room had a fireplace and was very pleasant with a sweeping view over the canyon to the mountains. The dining nook at one end looked down on the road scene we had just left. The kitchen was quite large with two sinks, a lot of cupboards and dominated by a huge black, wood-burning cooking stove, right in the middle of the room — for security purpose Alice explained. She called it the Black Beauty and all the earlier houses were provided with them, the sole means of cooking. I wondered how Army procurement had gotten hold of so many of these real museum pieces.

Mici Teller had a tea party that afternoon under a clump of pine trees. She explained in her picturesque speech — she was Hungarian — her victory over the Army, often interrupting her tale to fetch a cup for some passerby invited to join the party. "I told the soldier in his big plow to leave me please the tree so we could have shade. 'I got orders to level off everything so we can plant it,' said the soldier, which made no sense because it was planted already by wild nature and I liked it that way. The soldier left but was back next day and insisted he had more orders to finish this neck of the woods off. So I called all the ladies of the neighborhood to the danger and they came with their knitting and their children; we had a tea party under the tree, so what could he do? He shook his head and went away, and he hasn't come back." The Army personnel who ran the post had orders from Washington to go easy on their charges and to make things as pleasant as possible. When they were cornered on little features, they gave in. Fuss and feathers again for General Groves. Also I remember they thought that all the scientists would collapse and go a little bit nuts, so they were going to send a psychiatrist. After a while I learned it was the Army that had the worst of it. They were likely to leave and did. After dinner at the lodge that first day on the hill, we sat on the lodge porch, and looked out at the view, at the deepening twilight. I. I. Rabi, from New York, once remarked, "I envy all of you this magnificent view. None of you should complain about living here." He didn't live up there. Whereupon Emilio Segrè, who *did* live on the site and *did* complain, answered, "think how we shall hate, even the view, after living up here." But after two years, I never really got tired of it.

Los Alamos was a very unusual community. Beginning with its remote location high on the mesa, it was closed tight, exclusive to those chosen few possessing the magic pass. Only visitors or VIP's from high places in Washington, who were directly connected with the Manhattan Project, were allowed in. It was unusual because it was a young community, with an average age of around twenty-five years. There were practically no old people. Those of us in our early forties were the senior citizens. We had no invalids, no in-laws, no unemployed, no idle rich, no poor, and no jails. The architecture in Los Alamos was unusual too, a combination of log cabins and jerry-built Army construction, that sprang up everywhere, inside and outside the fence and in adjacent canyons. There were many log and adobe buildings from the old ranch school. Three large houses along what came to be called Bathtub Row (they had bathtubs) were divided into several apartments each. Two others were given to the Oppenheimers and to Commander (later Admiral) William Parsons and his family. Another large house I had thought I could have for school handcraft work was remodeled for the post commanding officer. Two of the smaller buildings behind the lodge were renovated for the British Mission. And smaller cabins near the commissary housed several post families. Across the field from the lodge was the Big House, a two-story dormitory where senior bachelors lived. One of the small stone buildings, former power houses, was made into one cozy room for George Kistiakowsky when he came up later. He complained that his heat was attached to the school and that since they didn't use or heat it on Saturday and Sunday he froze and had to go somewhere else.

The Army Engineers proceeded in the early spring of 1943 to erect barracks in the technical area. Barracks were built for the Army personnel and dormitories for the single civilians. They could eat in the mess hall near the Big House. Families of Army officers and scientists lived in duplex or four-family apartments alloted according to family size. Every Army building was painted a nondescript green and we soon referred to our dwellings as greenhouses. We had no sidewalks, no garages, no paved roads anywhere in town. We also had a trailer area for maintenance people at the edge of the mesa. The sanitary conditions were a disgrace and many of our socially minded ladies tried to get improvements, but in our last hectic days of '45, when the trailer area grew alarmingly, the authorities were too pressed to take any steps. Well, this then was the town where the first atomic bomb was made.

Plain, utilitarian, and quite ugly, but surrounded by some of the most spectacular scenery in America. We could gaze beyond the town, fenced in by steel wire, and watch the seasons come and go. The aspens turning gold in the fall, the dark evergreens, blizzards piling up snow in winter, the pale green of spring buds, and the dry desert wind whistling through the pines in summer. It was surely a touch of genius to establish our strange town on the mesa top, although many sensible people have very sensibly said that Los Alamos was a city that never should have been.

The strangest feature of all to us was the security. We were quite literally fenced in by a tall barbed wire barricade surrounding the entire site and patrolled along the outside by armed MPs. In our first weeks we heard shots but never knew why. Actually we felt cozy and safe, free from robbers and mountain bears. We never locked our doors. In our second year, extra MPs were sent to guard the homes of the Oppenheimers and Parsons, making round-the-clock patrols. No one, not even the families themselves could go in without a pass. If they had forgotten their pass, they had a hard time getting in. Some of the practical housewives cooked up a scheme to use these MPs as babysitters in the immediate neighborhood. What could be safer than a man with a gun guarding the precious small-fry? The children were sure to be impressed and behave accordingly. Martha Parsons never hired a babysitter as long as the MPs remained around her house, and Kitty Oppenheimer once got real service when the guard came to the front door of the house she was visiting to tell her that little Peter was crying. Soon after, the sergeant in charge put his foot down, no more babysitting for his crack MPs! a group that was specially picked for duty at the number-one government project. The patrol outside the fence soon ceased except for an occasional mounted patrol. There was little temptation to conquer the fence and no one tried, except dogs and children, to dig holes underneath it. Rather the fence became a symbol. We felt protected and very important, and tended to act accordingly, griping at everything, including our fenced-in condition. Although we could leave the mesa at will with a pass, we did have to keep within the boundaries roughly defined by Albuquerque, Cuba, Las Vegas and Lamy, all in New Mexico. We would go to Mesa Verde, Denver, Carlsbad Caverns or El Paso, with special permission. We could not talk to strangers or friends on trips and it was common knowledge that we were being watched by the Army G-2 and the FBI. In general, we were not allowed to send children to camp or away to school. If they were already in school they could not come up for vacations. Our driver's licenses had numbers instead of names and were not signed. All our occupations were listed 'Engineers' and our addresses as Post

Office Box 1663, Santa Fe. With gas rationing in effect, most of the traffic between Lamy and Santa Fe and Taos was ours. All in all it looked more than mysterious to the state police when we happened to be caught for a traffic violation. One day on the Taos road a caravan of Army cars carrying a group of Nobel Prize winners and Deans of science, all traveling under false names, was flagged down. When the officer asked the names of the occupants each refused as politely as possible to give it. Tell that to the judge retorted the police as he wrote out the summons, determined to teach the almighty Army a lesson. I'm sorry officer, ventured one of the men, we can't appear either. Finally the Army driver soothed the irate officer with the promise to take the summons to his commanding officer who would look after it. And it took this commanding officer and the governor of New Mexico to come to an understanding about this.

In the fall of 1943 the daily bulletin delivered by a soldier and thrust in the kitchen door suddenly announced that all mail, incoming and outgoing, would be censored. The announcement caused quite a stir and a number of questions about its fairness, necessity and legality. We were always accusing Army management of being dramatic about such things. As censorship began we had to apply for cards to send to relatives stating that mail was being opened for security purposes and asking that they destroy the cards and not mention the censorship ever. We sent our mail unsealed with the understanding that it would be read, sealed up and sent on. If something inside did not meet with the censor's approval, it would be returned to the writer with a slip indicating what rule had been broken. We each had a book of rules describing what not to say. We could not mention last names, give distances or places nearby, and the worst word, "physicist," was strictly forbidden. I might say we could write "theoretical" or "experimental," and the censors wouldn't know, but our friends would. The censors were human and funny things happened. Alice Smith returned a bill to an Eastern store saying enclosed please find my check. It quickly came back with a note from the censor, "lady, you forgot to put in the check." My son Jack, a stamp collector, once got a letter from a stamp company asking the meaning of the note enclosed in Jack's letter to the stamp company, and signed by the censor. I have the habit of decorating my epistles with pumpkin faces, with smiles up or smiles down, and the censor would send my letter back each time one silly face appeared. I argued and persisted to no avail and finally had to give up my sketches to get the mail through. Since Los Alamos, or more strictly speaking, P.O. Box 1663, was the only place in the United States where mail was censored, envelopes with the censor's seal are now collector's items. I destroyed

I don't know how many before we left for home, alas. We continued to live in a security-minded atmosphere for nearly three years. Actually, anyone who had wanted to could have given away secrets. But enough of us, while poking fun at the security regulations, took our trusted positions very seriously. Some of the neighborhood philosophers at Los Alamos foresaw implications in the secrecy formula. Hans Staub, who was a Swiss physicist, went around asking in emphatic tones of prophecy, "are these big tough MPs with their guns here to keep us in or to keep the rest of the world out? There is an important distinction here, and before I leave this place I would like to know the answer."

The U.S. Army Corps of Engineers Manhattan District ran the project in those early years. Most of us were civilians at Los Alamos. We found living in an Army post unique and I'm sure the Army regarded us, all of us, as equally strange. Ordinarily Army officers run any post to suit themselves, setting the standards and following strict protocol. At Los Alamos things did not work that way at all. Of all civilians, probably free-wheeling scientists with their tradition of non-conformity, are the least likely to measure up to proper military standards. Furthermore, there was a feeling that we were slumming it up there in our secluded mesa, far from city and university life and free from the need to keep up appearances. We truly believed in plain living and high thinking. To counteract the Army regime, the civilians had a town council, appointed at first and later elected as the town grew. The council was most unorthodox on an Army post but Oppenheimer believed that a civilian governing body, though lacking in real authority, would serve a useful purpose. And it did.

From the very first I was conscious of living under the Army. A soldier came to the kitchen door while we were eating breakfast to deposit the daily bulletin put out by the Army post. It was mimeographed, stapled together and bore the admonition — "this paper is for the site, keep it here." It had local announcements and was the Army's sole means of notifying the town's people of policies and regulations. Doc Barnett, in his Army uniform, came to call one of the first mornings after we arrived, when we were still slightly dizzy and nauseated from the altitude. Doc was young, just out of medical school, serious, and very good looking with blue eyes. We were not really ill, but he made notes of my boys' medical records and came to call every day as he made rounds of the neighborhood. He said he wanted to keep an eye on all his children to prevent real sickness. So many babies were born that the hospital had at one time nearly half its capacity used as a nursery. The whole town wanted to come and see the babies, especially when a little Oppenheimer was born. The sign 'Oppenheimer' was placed over baby Tony's crib and people filed by in the corridor for days to view the boss's baby girl. General Groves complained about the rapid increase in the population which immediately increased the housing problem and eventually would increase the school troubles. Rumor had it that the General ordered the commanding officer to do something about it. It is not clear what, if anything, was ever done. Our population was young and vigorous and the babies were free, so what could the General expect?

In our first year we had two academic scares, first, rabies in the dogs, and then polio. It seemed as though everyone had brought a dog to Los Alamos and with no security rules to bind *them*, they roamed over the mesa at will. When one of them began attacking people and was found to have rabies, Doc Barnett had a few hectic days. He kept bitten people under observation, gave extra shots, issued orders in the Bulletin and did his best to prevent the panic that was spreading fast. We were so frightened we obeyed all rules without question. When the dog owners got tired of keeping their pets inside or on a leash, they suggested putting the children on leashes and letting the dogs go free. Others began to talk of banning all dogs from the site in the interest of community health. People threatened to leave and it became a row of dogs versus children. Then almost on the heels of the first came the second epidemic scare. A young teacher was stricken with polio, and taken to Bruns Hospital after her serious illness could not be cared for at our small hospital. She died a few days later, spreading the deepest gloom over our small, close-knit community. School promptly closed. All children were ordered to remain indoors and not to visit each other. No one was allowed to go to Santa Fe. The commissary was quiet and peaceful without dogs or children. Snow began to fall and the children pressed their noses to the glass panes, longing to go out and play. A stillness of death descended on our mesa. Mothers had to stop work and stay home with the children. Little groups of adults gathered in back porches to hear further developments. Doc ordered mothers to report immediately any sign of fever or illness. The heart went out of everything in our community. There was no party of any kind for several weeks. Henry Barnett and Jim Nolan, both children doctors, bolstered our shattered spirits, performing far beyond the call of duty to dispel the doom and gloom. We had faith that the doctors would pull us through and they did.

In the early days, newcomers were conducted on a sightseeing tour around the commissary shelves. I was taken around on my second day on the site, before they had converted from an Army store to a supermarket. The first spectacular shelves contained rows and rows of gallon jugs of imitation vanilla or lemon flavoring, gallon jugs of jello powder, in all their delicious colors, and two-gallon jugs of tomatoes, peas or corn. You're in the Army now, it was said. But we also had products which had vanished from the shelves at home, chocolate, nuts, stuffed olives, tomato juice and some very fancy brandied dates, I remember. Those were never replaced. Vegetables, in those early days, were laid out on a sloping counter, lovingly presided over by Mr. Gonzales, who had worked for the Ranch School. He was a small man and so gracious and polite that he had difficulty in rationing anything in short supply, such as lettuce. He sprinkled his wares by hand and did his conscientious best to freshen up the tired vegetables. One of the crosses we had to bear the entire time was the milk situation. There was no delivery to houses and delivery to the commissary was uncertain and in quantities too small to go around. If an alert housewife saw the truck arrive she would shout the news to her neighbors and everybody would rush over to get there as fast as they could. Army procurement added more and more dairies to our supply, but our consumption increased faster. Mr. Gonzales, who was also in charge of the big, old-fashioned refrigerator, tried to ration the milk. He put very carefully lettered signs on the outside of the icebox — one quart to each person — but everyone claimed to be buying for others. The Spanish girls checking out groceries lost control over rationing discipline, but when they were replaced by WACs, *they* put us in our places. WACs are very good about that.

General Leslie Groves was the real head of the Manhattan District, but he did not live in Los Alamos. He was an absentee landlord, mysterious and unseen, and he got the blame for everything that went wrong. Mrs. Robert Wilson, Jane Wilson, complained to him that cooking at high altitudes on her Black Beauty was impossible. She invited him to come to dinner, which he did. Jane thought that perhaps the shipment of electric hot plates that was sent up later was the result of her dinner party. But we had many small benefits, courtesy of the Army. Soldiers cut and stacked wood outside our houses for our fireplaces and Black Beauties. Soldiers came with trucks and work gangs to collect garbage and trash and fix plumbing or anything else. At Christmas they went into the mountains and cut trees of every size for us to choose from. Heating was taken care of by the Army, which seemed like easy living to those of us used to stoking our own furnaces. But these furnaces were very special. The first stoking of furnaces came to be a sort of Fall ritual on the mesa. Cold came early at our altitude and by the last of September a frosty chill crept down from the Jemez Mountains, turning the Aspens golden. Indians were taking in their strings of chili peppers and corn and gathering

piñon nuts for Winter. And we went down to the valley for apples from the
orchards. At about this time a notice appeared in the daily Bulletin. "In the
near future furnaces will be put into operation. At that time it is expected
that there will be a small amount of dust blown into the apartments. Inas-
much as there is no way of removing dust from the ducts, your indulgence is
required." We discovered in our first year that this was a studied understate-
ment. So the second year, remembering the clouds of soot emitted near the
opening near the ceiling, we placed cheesecloth over the ducts to catch the
worst of it. The windy blasts were apt to blow off the cloth. After some days
the soot content was slightly reduced, but never to zero. Our furnaces would
never have passed any city inspection, but then we had no city inspection.
Each day before dawn, during the furnace season, the crew of stokers would
arrive. They were an unworried Spanish and Indian lot who sang and beat
rhythms on drums or cans, and drank Cokes in the warm furnace room in the
basement. We heard it all through the open ducts in the wee hours of the
morning, before anyone was fully awake. Then they finally left and we would
fall asleep again only to be aroused by a sound like a tremendous waterfall
followed by the rush of hot air mixed with cinders. In the dining room at
breakfast the thermometer might register 90 degrees, or quite often 50
degrees. Changing the thermostat had no effect whatsoever, much to the
bewilderment of our very mechanical-minded family. The Army admonished
us to let the furnaces strictly alone. In emergencies the approved procedure
was to notify the major in charge and lodge a complaint, whereupon the
major would relay the complaint in English to the Spanish speaking stoker
gang who would presumably take care of the problem on the next tour of the
furnaces. But our emergencies didn't always lend themselves to the system.
In our four-family house an emergency, by common consent, was when the
inside walls sizzled when touched with a wet finger. A complaint would be
fully lodged with the major, part of protocol. Then while the chain of command began some husband in our house, usually mine or Cyril Smith, would
boldly march down with fire in his eyes to the forbidden but unlocked furnace room and, as everyone watched from the window, seize a shovel, remove
a mound of red-hot coals and dump them on the side of the road. This at
once brought showers of congratulations for the brave men and reduced the
heat so much we were glad to welcome the picturesque stokers on their very
next visit. Like the Black Beauties in our kitchens, the furnaces were museum
pieces with their function further complicated by the uncertain ability of the
stokers, and the inferior quality of the coal which came from Madrid, near
Santa Fe, and was so poor it could not be used in any other part of the war

effort. That's why we had so much of it! A huge pile of coal brought in all the time by big trucks for the entire project, lay between our green house and the next green house. Trucks came in winter, *all* hours of the day, hauling coal to all parts of the mesa from there. Coal dust was one of our neighborhood ingredients, tracked inside by children and their pets. Of course the little kids loved the hunks of coal and bits of cinders and dust so readily available. I once saw two little toddlers living in the lower part of our house sucking coals, just *covered* with black dust. Well, their mommas went out and grabbed them up and put them in the big sink. One of the sinks was deep, so you could bathe babies in it.

Besides the regular Corps of Engineers who ran the post, a special engineering detachment was sent out to work in the tech area. They were quite different from the regular post soldier. They looked, in spite of the uniforms, like budding professors instead of combat troops. Shortly after they came up to the hill some high brass from Washington came for a formal military review in the baseball field in front of the Big House. All of us came with our children to see the show. The MPs, the post soldiers, the WACs and even the doctors made a fine upright showing as they marched across the field to band music, but the newly arrived SED boys were terrible. They couldn't keep in step, their lines were crooked, they didn't stand properly, they wore glasses and they waved at friends and grinned. Well, the situation was not helped by the fact that they received a lot of applause from the bleachers. The visiting brass let it be known, when they went back to Washington, that the whole post was a terrible disgrace to the Army. But the SED boys worked long hours in the tech area. Although they often worked late into the night to meet deadlines, they were expected to arise at dawn for inspection and drill by tough sergeants from the regular Army. Once when a sergeant became irritated by his yawning, half-hearted crew, he shouted, "if you guys think I like this job, you come up here and try it." One of the SED boys offered to lead the drill in his place. He shouted orders in imitation of the sergeant's voice. "Thumbs up, thumbs down, wiggle, waggle." Even the sergeant broke down and dismissed them. My husband and others who used the SED boys finally got the discipline relaxed. The drills stopped and bed inspection let go so they could sleep in the mornings. We became accustomed to administrators and doctors in uniform; WACs selling sodas and checking groceries, selling postage stamps and cashing checks; and military police with guns guarding fences and gates and keeping our comings and goings under strict, watchful eyes. Perhaps it gave us a sense of being a part of the war effort. It certainly helped Santa Fe people to believe that we really *were* a war project. It was said that they had their doubts.

The technical area, called T-area, Tech Area or just simply T, where the main work of the project was done, resembled a small factory – a two-storey clapboard building painted green, of course. The windows were large and pleasant, like those in our houses, although innocent of any washing since the original putty was smeared on. This one building designed as a laboratory only was built along the west road, but it soon grew in several directions and added wings whenever possible.

The physicists were divided, roughly speaking, into two varieties, the theoretical and the experimental. The distinction often made among themselves was that the former knew what was the matter with the doorbell while the latter also knew how to fix it. (Whether they did fix it is something else altogether.) Everyone wore causal clothes, jeans or old unpressed trousers, open shirts and no ties. I don't recall seeing a shined pair of shoes during working hours. They all seemed to be enjoying themselves as scientists always do when they ponder their problems together. No one has to drive them; they drive themselves when they have an intriguing problem. And so it was at Los Alamos. Even an outsider like myself, with no idea what the problem was, could feel the inner urge for scientific solution.

I suppose I heard a lot of talk which is even now stamped top secret, and I used to ask facetious questions when the talk seemed to get bigger than usual. I once asked Emilio Segrè what on earth we were hatching up there? To put me in my place he answered, "now, just you listen to me, what we do here, if we do it, will make a revolution, like electricity did." I knew we were engaged in an important aspect of the war effort, but as to Emilio's revolution, I continued to discount it. But later I suppose he was proved right. Each night after everyone had gone home, the MPs came into the T area to check out security violations. If they found anything, they would simply write an appeal in the Bulletin next day, for more diligence. But once in a while they cracked down. One night, after midnight, Hans Bethe was selected as their victlm. He had left something out of the safe, so two MPs came to his house, woke him up and insisted he return to T and put the stuff away, himself, to teach him a lesson. Everyone was more careful from then on.

The usual part-time percentage for working wives such as myself was 3/8ths, never one-half, or three-quarters. Why it was 3/8ths remained one of our favorite top secrets. When a notice appeared in the Bulletin asking for applicants for this 3/8ths time job several of us wives applied. Although none of us were trained in science or maths, they needed help and we were accepted. We were trained as computers by Joe Hirschfelder. He was a chemistry professor and ballistics expert, and had an office, a small office, and a great big room where we worked. In his office he tacked up Esquire calendars, as he said, "for decoration and cheerfulness." His secretary was a Santa Fe girl in tailored clothes, her hairdo in perfect order and wearing stockings and high heels. She told us, eyeing our sweat shirts, jeans and sandals, that she worked for a bank in Santa Fe before coming to the site, "where everything was of the best, Mrs. Brode. You could not run a bank like this." She was not used to the untidy professors, and said she could never bring any order out of Joe's desk. Joe said she was a pack-rat and always wanted to put away his stuff.

Joe was a bachelor at that time and insisted he be allowed to have his mother come up and keep house for him. He told Vera at the housing office that his mother was advanced in years and should have a lower apartment, usually reserved for families with young children, so she'd not have to climb stairs. When she finally arrived, on her first day, she took a lengthy hike up three miles of steep trail, to "see the splendid view, don't you know?" Mrs. Hirschfelder, or Ma, as we affectionately called her, was short and stout, with a ruddy complexion. She wore black dresses, high black shoes and a large black hat with Pink roses, and could be seen walking everywhere, with a black umbrella in bad weather. She was an enthusiastic supporter of all activities and thought everyone and everybody was most amusing. She wouldn't have missed living on the mesa for anything.

After three months of computer training, some of us moved to the theoretical wing of T to work under mathematician Moll Flanders. After hours, Professor Donald Flanders and his family contributed greatly to our musical life. All the family sang and played instruments and Moll himself conducted our chorus and orchestra. He wore an unusual, but becoming full beard. Other people didn't wear beards in those days and when we asked why, he replied, in his New England accent, "well, if you had three women in the family and only one bathroom, would you not do likewise?" Moll put us all in one large room with the machines around the edges and our desks pushed together in the center. Now, this was a mistake because we all sat facing each other and it was a temptation to discuss mesa business, in which we all were more knowledgable than the work at hand. Our immediate overseer was Mary Frankel, a young wife who set up the problems for us to run through the machine. The resultant figures were taken into the graph room in charge of Bob Davis, and were plotted on graphs. The idea was to make curves and all the scientists took the greatest interest in the progress of the curve. If our figures put a jog in one of these curves, our figures were wrong, not vice versa. Mary had a small office near Moll's where she made out the problem sheets and put them in a basket marked "Free — take one." She was a young blond with a Ph.D. who took her job very seriously and encouraged us to study maths at night. I don't think we did. She was fussy about decimals and regarded our mistakes as appalling, which they probably *were*. "Mrs. B.," she would say, "there's a great future in this computing business if you could possibly learn decimals." Mary was a good twenty years my junior, so I reminded her that *this* was my future. I once asked her why we worked 3/8ths time and she replied, "Mrs. B., if I could answer questions like that, well, I'd be in charge of this whole outfit."

Now the wing of the technical area where my husband, an experimental physicist, worked was at the other end of the original T building. It was here that still top-secret problems were accomplished. This was the noisy part of the technical area, with the hum of machinery sometimes shaking the floor. The large labs adjoining the offices were full of enormous gadgets labeled with crude warning signs like 'Caution – live wires – test well before using', 'Watch light', 'Keep out', 'Don't handle – wear lead'. I was most careful where I stepped and I never tried to sit. All the chairs were apt to be holding sharp tools or bits and pieces and 'parts'. Long rolls of paper were emitted by jerks from a machine and the men tending all the machines had that characteristic rapt expression. The din was terrible. People had to shout to be heard.

When the WACs arrived, well-trained as computers, wives were permitted to quit and only a few of the wives without children stayed on. I was allowed to keep my orange badge, however, so I could get into the T area and run errands or pick up our mail. We had no mail delivery to our homes and when my husband was off the mesa, which he frequently was, I could get no mail. There was also a certain prestige involved in possessing a badge admitting one to the inner sanctum. Many persons in our town lived all through the war years without a peek inside the T area.

Salaries were paid according to the last job held, resulting in many inequities. But this was accepted as a sacrifice of the war. There was no time to work out another salary scale for everyone. Young men from lucrative jobs in industry and hired for routine work on the project were often paid handsomely, while senior professors from the universities who bore much of the burden of the project received much less. Young men without Ph.D.'s often contributed ideas but got stipends no more than meager fellowship grants. To compensate, we were charged rents according to salary, and not by rank or by size of the house. We were all cut down to size at Los Alamos and it was sometimes a sobering experience. Even Oppenheimer abandoned his former pattern of living. I remember the old days in Berkeley, when he would not accept a class before eleven in the morning so he could feel free to stay up late for parties, music, or ideas. But at Los Alamos, when the whistle blew at 7:30 — we had a factory whistle — Oppy would be on his way to T and hardly anyone could beat him to it. When Sam Allison came to the site from Chicago he shared Oppy's office for some time. Sam said his one ambition was to be sitting at his desk when Oppy opened the door.

Our school in Los Alamos was the most exclusive in the whole country, and our worst problem. In the beginning everyone was completely unrealistic about the school and we all had to learn the hard way. The earliest organizers seemed to proceed on the assumption that the school would be the least of their troubles. With all those thoroughly educated people to be hired, why wouldn't it be a cinch to set up and run a fine school for their kids? I shared this smug optimism myself. Here was a chance of a lifetime to put our own ideas on education into practice. When I heard that we of the new community could run our own school, I had visions of my boys learning perfect French from Peg Bainbridge, who had taught at Radcliffe; with some of the country's finest chemists teaching my Bill the advanced work which he really was ready for. Take any subject and there'd be some expert in that field and he could teach our children. To get off to a grand start, the Manhattan District hired a midwestern educator to make a survey. The survey was kept top-secret for some time, but we got hold of a copy about the second year. It revealed the concept of the super-school, with super-children and super-parents, all adding up to super-education. And I imagine the educator planned to follow it up with a future study of the super-results. The survey laid down plans for a super-building too. And *this* we had. It was the only super-building on the mesa. Everything else was jerry-built, but not the school, and I *never* have found out just why. The school was modern and exceedingly well built to last 100 years, whereas the technical area houses were jerry-built with no assurance they'd even last out the war. We loved our beautiful school building, with its picture windows looking out over a super view, and we used it for meetings and parties as well as classes.

Rumor had it that the General was a bit less than pleased at the original construction. According to the story, the original plans called for a foundation of solid cement on the hillside. When the location chosen for the school was found to be solid rock, the Army blasted it out and filled it with cement, according to specifications. One day as General Groves was touring around the mesa he inquired how the school was getting along. When he saw how, he fired the poor, unfortunate major right off the mesa. A count was made of children, apparently, but by slide rules, statistics of population, mean tests and whatnot. No one came to the site. One man said he did come to the site that summer, but I never met him. He said he came to count noses, but he didn't take any consideration of the *age* of the noses, which was important. The average age at Los Alamos was twenty-five so the children were babies, or yet unborn. The overcrowded nursery school was not included in the plans and had to be privately organized by the mesa mothers. Added to the miscount was the underlying assumption that the entire community would be composed of brilliant scientists who would not only have brilliant wives, but children with abnormally high IQ's. This was *not* in accord with the facts. Only eight of the forty students in the upper school, that's from sixth grade through high school, were children of staff members, and none of them were really above average although they all expected to go to college. The rest of them were mostly Spanish speaking. When school opened late in the fall none of them had a mind to settle down into formal classes five days a week. All the teenagers had roamed the mesa the entire spring and summer and well into the fall when the school was being built. The Army had been very generous to them, lending them trucks and jeeps for trips, and got mad only once when some of the kids deliberately pushed a jeep down the side of a steep canyon for laughs. The kids were only scolded and told they could have no more jeeps for a while. These children sensed at once that the school was new and the management uncertain and they made the most of it. They were not bad kids except for one or two who led the pack and bragged about having been expelled from several schools in New Mexico. These few wielded enough influence on the group to take them all down to Santa Fe on a school day in an Army truck. I remember Alice Smith poking her head in my door on her way home at noon, saying, "all gone to Santa Fe again, my class is only seven — staff member kids — Bill and Jack Brode, two Flanders girls, Joanna Jorgensen, Jane Kline, Kay Froman and Lloyd Williams." Even the regular Army bus would stop long enough in Española for the teenagers to run into liquor stores for whiskey which was not available in Los Alamos. Some of us tried to organize parties for the kids before school opened but the gang would break up anything.

The lower school did not have the extreme disciplinary problems of the upper school, but it had minor unusual problems such as lice on the heads of the first grade, which disrupted the routine for weeks. Doc Barnett went up to school every morning to go over the heads, including that of the teacher, with beautiful red bushy hair. Doc taught the mothers how to hunt for lice every evening and he ordered all heads shorn as short as possible. My boys looked very queer. Crew haircuts in those days were associated only with prisoners of war. Mothers of little girls with golden curls wept when the locks were cut. We also had a bout with ringworm at the same time.

Another prediction in the survey was that no extra teachers would have to be hired. It proposed that *wives* of the scientists and post employees would do the teaching. As far as I know, none of the wives were queried about qualifications or willingness. Fortunately a few *were* well qualified. Alice Smith agreed to manage all social science. Jane Wilson was a graduate in English and offered to try. Betty Inglis was experienced in teaching maths and Mrs. Long was a chemist with teaching experience. She was the one who died of polio later. No one was available to teach any language during our three years, although we had natives from half the countries of Europe living up there and plenty of Spanish speaking personnel. Several wives, like Peg Bainbridge, had Ph.D.'s in language, but claimed they knew no grammar and had held only graduate seminars in the colleges. Furthermore, both Peg and Françoise Ulam had babies to look after. But we did prevail upon them to have small groups in French and Latin, after school. Except for a young technician sent over by the tech area at one time, we were never able to get any physics or any chemistry taught. The second year we had a teacher from the outside who gave some biology. It was not until after August 6, 1945, when all fathers of the school and some not-fathers agreed to come and give one lecture apiece, that we had any physics. Our professors included Fermi, Peierls, Williams, Brode, Mack, Froman, Jorgensen, Smith and Feynman.

Soon after Christmas the first year, Betty Inglis walked out of the school after the limburger cheese incident, and refused to return. One thing after another went wrong with her class. She tried to be very disciplinarian, 'till that fateful Monday morning when she opened her desk drawer and found it filled with limburger cheese left there since the Friday before. When the incident was related to Oppenheimer he said, "that school does stink doesn't it?" Alice and Jane stuck to it until the end of the war, when we all went home. They deserve the greatest credit, for their responsibilities grew with increased enrollment and extracurricular activities. When Betty could not be replaced, as a last resort, I was paged over the intercom at T to come to the school, being asked to hold down the maths class. Although he was in a fine position to know my mathematical shortcomings, Moll, my boss, said that I'd better go. So I did, and his eldest daughter, Ellen, and Lloyd Williams really conducted the class. The first superintendent sent up by the survey people left after a short time. It was quite a while before we could get another since all applicants had to be interviewed off the mesa, investigated, and told just enough to entice them but not enough to frighten them. Meanwhile the lower

school principal, who also had the third grade, assumed the entire responsibility and threatened to leave unless she got help soon. The upper school by
this time was completely out of control. The authorities reluctantly agreed
to hire another teacher who could give subjects like typing, art, and elementary biology.

We got a friend of ours, Marjorie Crouch, to come up and try her hand.
Marjorie was a tough disciplinarian and had been successful in other jobs. I
did not share many of her ideas on education, but by this time our fancy
views seemed beside the point. After she was interviewed and agreed to come,
I wangled one of the single apartments for her, reluctantly given for one
person. I was tired of our only school being treated as a step-child by the
Manhattan District. As soon as she arrived we had dinner at our house and
then took her to her apartment. No sooner had she gotten inside the door
than the entire school board called on her. They were a bunch of desperate
men who promised to back her up on any measure she thought they should
take. The latest incident had been the shooting of a large school window full
of bullet holes. The culprits bragged about the feat, but the Army held the
school authorities responsible and refused to discipline the boys. One of the
school board members, Johnny Williams, said "next time you put in glass,
put in bullet-proof glass." When the school board took Margy to visit the
school that evening she admired the view of the snow covered mountains
through the bullet holes. Well, Marjorie took charge the next day. As soon as
she was introduced, "this is Mrs. Crouch, the new principal," Miss Saunders,
the former principal, almost ran out of the room to her nice, little third
grade. Marjorie cut short her planned speech and instead asked if there were
any questions. After an embarrassed silence, Lloyd Williams rose to his lanky
height. He was a handsome lad of sixteen and assumed the role of spokesman
for the gang. "What exactly will be your policy on unexcused absences, Mrs.
Crouch?" "Well," answered the new principal, "since you must have a particular reason for asking (everybody laughed) you might explain further what
you have in mind." Lloyd was equal to the task and explained how all employees in the tech area were allowed one half day a week to go shopping in
Santa Fe. They all worked a six-day week. So, he argued, why shouldn't the
school have a similar privilege? Marjorie remembered the tales we told about
hooky and she caught the six-day week phrase in Lloyd's presentation. "Well,
I don't know what my policy will be, but it will be adequate," and it was.
Saturday school was instituted. Poor Alice and Jane quaked at the idea and
predicted no one would show up on Saturday morning. On the following
Friday, Marjorie read off names of those who had been absent during the

week and would be 'eligible' for Saturday school. Everyone eligible turned up and each was given a theme to write all morning. It was good hard work and no fun at all. After several Saturdays of this, unexcused absences ceased and playing hooky became a thing of the past. Little by little the climate changed around the school and the children began to enjoy it. Even their manners changed. They took gum out of their mouths before reciting, wore less sloppy or dirty clothes, and sat up straight instead of lounging with their feet on the desk. Marjorie made a few enemies among the post authorities, the school board, and even her best friends, but on the whole we all supported her and were eternally grateful for her efforts.

The school did improve and it did serve the majority of the students, but it never became the dream school we expected for our children. But we got the whole upper school under control and that was the main problem. Then we had a new superintendent, another one, Esther Swenson, and she and Marjorie obtained results we had not thought possible. When the town population increased, bringing many more upper school children, it was a mercy the school had come under constructive management.

The Indians were brought to the site by the Army every weekday, including Saturday, and they would work for us for half days. They were alloted to us according to our jobs of importance. If you worked 3/8ths time, you could only have a half day's labor. If you worked full time you could have two half days. They were very picturesque as they trudged up our road. After I became acquainted with some of the Indians I learned that they liked us on the hill because of this working relationship. It was the first time they had known any group of Anglos who were not primarily interested in their welfare or curious about their cultural patterns. We accepted them as a part of our strange life on the mesa. We began to visit their ceremonials and fiestas like any other tourist, but before long we went to the pueblos as their guests. Many persons also became acquainted with Spanish families in the same fashion. I first met Po, the son of Maria the pottery maker, at the Sadie Hawkins Day party, when the square dancers put on a demonstration and we were waiting backstage in the wings to go on. Po was waiting too, dressed in feathers and paint for his war dance number and looked very glamorous and wild. His dance was even more so. I'd never seen an Indian dancer before and later I discovered that he had never seen a square dancer, so we smiled shyly at each other. I used to meet him in the corridors of the tech area, both of us dressed in drab workclothes, but we recognized one another and always smiled. Once I stopped him to ask why he didn't come to our Saturday dance group and he promised he would. He did, with his pretty part-Spanish wife

Anita, and they learned to do the square figures with very little help. Anita and I talked about religion and how the Indians felt about the dual system. They are all good Catholics and at the same time they have their traditional gods and ceremonies. This seemed very strange to us, to watch the corn dancers take out the Catholic saint statue from the church and place it in a shrine up for the day, then perform their ancient heathen dances which had no relation whatever to Christianity. When the dancers stopped to rest they had a choice of going into the Kiva, down below the Indian ground, or kneeling in the shrine to pray before the saint. At sunset they took the saint back to the church, the dancers still in corn dance or war dance costumes. Anita saw nothing strange in this. It had always been so since the Spanish conquistadors forcibly converted the conquered Indians to Catholicism.

The Indian dances, whatever else they meant to the Indians, provided good fun and a show for everyone. Our people at Los Alamos provided a good and enthusiastic audience and the Indians liked it. Some of the dances were held on weekdays, but when it was understood that our men would come only on their one day off, Sunday, some of the dances were switched to Sundays. The dances also increased in variety, we noticed, and we suspected the Indians were making up new, non-traditional ones since we were enthusiastic but uncritical and would not know the difference. We heard tales from Santa Fe people who knew the Indians well that the influence of Los Alamos was deplorable, but we actually encouraged the Indians to break away from some traditions. Most of us felt that many Indians were tired of being pressured to remain highly traditional, that it was for the benefit of the tourists and even their well-wishers. Some of the young Anglos from the hill, our young men, put in electricity for San Ildefonso and soon refrigerators and appliances made their appearance in the pueblo. Such improvements were not artistic or romantic but they made life easier. We were rather shocked ourselves to find Grand Rapids furniture, brass bedsteads, linoleum floors, soda pop and ordinary dishes in Indian houses. Some of us had more Indian crafts in our Army apartments than the Indians had in their homes. I had a carved table and a set of Maria's black plates and candlesticks. Maria herself set her table with a store tablecloth and store dishes on it. I once saw in a corner bureau in her front room new pink pottery mugs and a picture of the kind you bought in any cheap store with a card tied on it saying 'Happy Birthday to Maria.' I looked across the room to her display of black and terra cotta pottery for sale, laid out on lavender oilcloth and wondered how this could be. Well, most of the Indian houses had walled up the fireplaces, picturesque fireplaces where you used to put logs, and they used smelly kerosene stoves. Maria loved her pottery, even if she did not eat from it.

In our last year, Marjorie Crouch, the schoolteacher/principal, arranged for Maria to come to our school. Very early one morning Marjorie went to San Ildefonso to fetch Maria and all her pottery makings, including extra clay for the children. Maria squatted on the floor at the school and showed how each step of the process was done. She didn't say a word. Actually she speaks very little English. She held their attention for hours. Her quiet charm and dignity as she worked carried over to the children as no lecture could have done. She coiled her pots and smoothed them with small stones and her thumb. The Pueblo Indians do not use a pottery wheel. She gave each child some clay and took their work home to fire in her next batch. In due time her son Po brought the children's pots back.

The grand finale of our association with the Indians took place on a cold December night in 1945, when the square dance group from the hill was invited to a party at San Ildefonso. It was a joint affair arranged in detail by Po and a committee from our dance organization. It was Po's brainchild to celebrate the atomic age in general and the new relationship with their Anglo-scientific friends in particular. Our committee went down to the pueblo one afternoon to look at an old recreation hall built by the Indian Service, but long in disuse. It was big and bare and very dreary. Po insisted it could be fixed up, the roof mended, decorated and a stove put in. We then went to confer with Maria, who apparently kept close tabs on pueblo affairs. She had strong feelings against liquor of any sort, even beer, and she took me aside to make me promise none of our people would bring down any liquor. I promised on my honor, knowing how difficult this might be to accomplish with some of my gay colleagues. I went the rounds personally to impress on any doubters that Maria said the party would be spoiled if any of us arrived with liquor, inside or out. This did not add to my popularity and I felt like everybody's grandmother and a kill-joy. We brought several cases of Cokes, so dear to the hearts of the Indians, and Anita provided fruit punch and pitchers of water. When we arrived we saw a dais had been constructed against the wall. Po got up to make a speech of welcome, first in English and then in his native Tiwa tongue. Whether he made the same speech in both languages I have no way of knowing — there may not be a word for atomic nor for scientist in Tiwa. The British scientists in our group whispered that Po resembled exactly an English country squire opening up a bazaar. He certainly had the air of lord of the manor. He had been too, for he had made both sides of the pueblo, previously a little divided, unite for the party, and worked hard to get the hall and food ready.

Four Indians opened festivities with a Comanche war dance accompanied by drums and a chanting chorus of men led by Montoya, our janitor from the Big House. Next Po called on us to put on a demonstration of a square dance. We formed four squares which we had practiced especially with four callers: George Hillhouse, Mat Sands, Bill Elmore, and one SED boy. We used our most seasoned dancers to make a big impression. Most Indians had never seen square dancing before. After we finished we asked Po to invite everyone to join in. We mixed the formation so some Indians were in every square along with our experts. They were natural dancers; even the fat women were light on their feet and had perfect timing and rhythm. By this time the initial shyness had disappeared. They were all laughing and having fun. After a short interval some of the Indian men, Cokes in hand, began to shuffle to the drums that had started up. They tied the blankets which had been over their shoulders around their waists, took hold of some of us and indicated that we should shuffle around with them. Our drinks finished, Po shouted in Tiwa the directions, which we gathered were for a sort of serpentine styled dance game. The old governor led out with his blue and white checked blanket, tightening his bobbed hair for dancing and his grin showing two front teeth missing. His moccasins kept perfect time, and he made gyrations with his arms which we were supposed to follow. Every once in a while the serpentine line broke and we danced like a Paul Jones with whoever was nearby. At the height of this excitement, with yells and shouts, Montoya, who was playing the drum, got up on a chair and shouted above the din, "This is the atomic age, this is the atomic age!"

Then the drums ceased and we flopped into the nearest chairs and onto the floor to recover. This was the intermission for eating. A long table, running the entire length of the hall, was set out with plates of food, all looking unreal and exotic, nothing we had encountered before in Indian feasts. The pueblo wives, under prodding from Po, had dug up ancient recipes and produced almost forgotten pastries, hot baked dishes, lots of chili, fancy squash mixtures, prune pies, like pemican cakes, and tiny tamales. We'd never seen that food before. The Indians just ate whatever they could get at the Commissary. There were strange things we could never figure out. All delicious and our hosts seemed to be enjoying the food as much as we were. There was plenty of coffee and pitchers of fruit juice. The guests from the hill were invited to eat first. I noticed the Indians held back so I spoke to Po and I said we wanted no segregation so he gave orders in Tiwa and everybody joined in.

So we took plates and everybody helped themselves. I noticed that they all lost their famous inscrutable expressions. It was a wonderful evening, but by 2 a.m. we were tired out and thought we should not wear out our welcome. Besides there was the long drive home. Anita and Po seemed to feel something was wrong that we would not stay longer and they danced on and on.

No one planned any deliberate association of scientists and Indians but undoubtedly life at Los Alamos had its effect on the pueblo. For one thing, their reservation fields are still uncultivated. They worked for wages on the hill and bought food in the stores. I got the impression in 1948, when I visited the pueblos again, that the Indians were eating far better. They certainly had new pieces of furniture, new appliances and a Spanish household near Española had new wings. I stayed with Po and Anita that time and she pointed out, in her new house with a bathroom inside, that there were two beds in the bedroom. She had bought them when the British Mission head, Sir James Chadwick, was leaving the site for Washington. Lady Chadwick had sold their furniture which they had brought up there, and the Indians had bought the two beds. So I slept in Lady Chadwick's bed in the San Ildefonso Pueblo in 1948.

No saga of wartime Los Alamos would be complete without including our relationship with Santa Fe, the long-suffering city that bore some of the brunt of our super-secret mushroom growth over on the Pajarito Plateau forty miles away. The Santa Fe people tried to retain its picturesque reputation and to resist expansion. It is not surprising that the caravans of shabby cars and large Army trucks mysteriously turning up the road to the Pojoaque Valley and Española caused alarm among the citizens. In wartime no community can seriously object to Army posts being stationed in their vicinity. The unusual secrecy precautions, including the FBI and G-2 agents, the fact that no names were known of the newcomers, nor purposes of the new venture hinted at, and not even the exact location was revealed, not only annoyed Santa Fe but aroused grave suspicions. Anyone who ventured up the old Otowi road to see for himself where the caravans were heading was turned back at the East Gate by armed MPs. The new people gave only one address and we didn't talk to anybody. We were not allowed to fraternize with anybody. The influx of hundreds of new people into the area did cause hardships, especially for things in short supply. The local people complained we were using up all the laundry facilities, buying up parts for cars, hardware, liquor and so forth. Now, the shops and garages themselves prospered under our business, as we were about the only tourists. The main worry seemed to be that we would change the atmosphere in the ancient sleepy town. We may have livened it up a bit with our big shopping tours, but we had no effect whatsoever on the picturesque atmosphere well-established in Santa Fe. In those war years we often walked around the winding streets of old Santa Fe and looked enviously over the walls into romantic and exotic gardens, adobe houses and chapels within, and we couldn't go in at all.

Sooner than many expected, in August 1945, the news was out and parts of the three-year jigsaw puzzle fell into place. Santa Fe buzzed with excitement and now-it-can-be-told satisfaction. That city had a right to feel they shared in the birth of the atomic era. When the Army secrecy ban was lifted in December '45, we had a grand debut into Santa Fe society. It was staged by a special committee of Santa Fe citizens and held at the Museum of Anthropology. The Museum was divested of its Indian exhibits and replaced with pictures of atomic experiments and Hiroshima damage. Enrico Fermi had a demonstration set up and explained it to groups gathered around. A movie of Hiroshima was given two showings in a lecture room to a capacity crowd. The main event was a program of speeches by several of our scientists, introduced by the Santa Fe committee. Vicky Weisskopf and Philip Morrison made the chief talks, standing on the stairs of the large hallway. Phil had just returned from Hiroshima and described the damage in detailed graphic form. Both men stressed the fact that the A-bomb was made by representatives of many nationalities and pressed the hope that some form of international control would be the result. After the lectures a question period followed. Vicky called on a different scientist to answer each question, and each introduced himself and told where he came from before he answered the question. Sometimes a question could not be answered for reasons of secrecy. The audience was most interested to see the secret people and to hear their opinions about the hopes and dangers of the new era. After this serious part of the evening there was social hour. We didn't know any of them and they didn't know us. An old lady standing next to me said, "I want to know some of those scientists; where are they, I want to meet them." So we all began to introduce ourselves and that started our parties. They all wanted to have parties, so we were all invited. They would send invitations to our site: "Mrs. So and So would like to have 'so many' scientists and their wives" — it didn't matter who, they didn't know. And so, in a little room, I helped the boys to establish the Atomic Scientists Association, and we would put up a notice giving details of the party in Santa Fe, asking anyone who cared to go to sign up. So we all had a chance. At first we accepted with alacrity the many invitations that gave us a chance to see some of the romantic houses and meet those renowned people, writers and artists and anthropologists. We wives were the most thrilled, but of course it was the men the citizens wanted to meet and talk to. We had difficulty with our men. After one or two parties they balked. It took too much time for the eighty mile round trip in cold weather. Our good husbands were not too good at being lionized. It embarrassed them and they made no visible effort to improve. As a matter of fact they gave out completely, long before we wives were satisfied.

NORRIS BRADBURY

LOS ALAMOS – THE FIRST 25 YEARS

Though I had taught at Stanford, I had also been a member of the Naval Reserve, and before the war I was called up and into a blue uniform. By a circuitous chain of circumstances, I was assigned to Los Alamos with a rather small group of naval officers in 1944. I worked with Professor Kistiakowsky on some of the aspects of the implosion problem. When the first weapon was tested at Alamogordo on July 16, 1945, it was my responsibility to superintend putting the thing together and getting it up on top of the tower. It was also my responsibility to play some role in the preparation of things for their use in Japan.

At the end of the war, in early September of 1945, Dr. Oppenheimer very much wanted to leave. He'd had a frightfully hard career at Los Alamos. Those of us who knew him there, who worked with him, saw the strain under which he labored constantly; his dedication to his task; his incredible ability to pool very diverse groups of people together, being as diverse as one scientist from another, and as diverse as civilian scientists and military personnel; his extraordinary knowledge of every phase of the work, and his ability to contribute to every phase of it. All these left their marks physically and turned to deep fatigue. I mention this only because serious questions as to his loyalty were raised during the hydrogen bomb controversy. I might as well say right now that I never had such questions. I never saw any reason to entertain such doubts. I never saw a man in my life more fanatically dedicated to his country. And one can be dedicated to one's country without agreeing with everything his country does. At any event, in 1945 he wanted out and all of us who were there were in sympathy.

By a series of circumstances, I was asked to take over the job, partly because I had some experience beyond gas ions at Stanford, with the small cyclotron there, and partly because I had a good deal of experience in military ordnance, and I guess I looked like someone they might hand it to for a while. I agreed that I would take it for six months or until they could find a permanent director, whichever came sooner. That six months stretched into 25 years and it's still my home now for more than 30 years. It was one of my trepidations, I must say, that in taking the responsibility I would never get out of it. This turned out to be true, although I have never had the slightest regret of the decision I made or the task I undertook.

161

L. Badash, J. O. Hirschfelder and H. P. Broida (eds.), Reminiscences of Los Alamos 1943–1945, 161–175.
Copyright © 1980 by D. Reidel Publishing Company.

So in September of 1945, with the war recently over, Los Alamos consisted of about 5000 people. Of them, perhaps 2000 to 3000 were GIs and military personnel and 2000 civilians. The laboratory itself had about 4000 people working in it; the other thousand included the wives and the children. Many of the wives worked as secretaries or technicians or as professional scientists, depending upon their backgrounds. The GIs were mostly what we called the engineering wonders: people with two or three years of quick training in engineering or science — physics, chemistry or mathematics — put in uniform and sent to Los Alamos. Some were Ph.D.s; I had a couple.

It's now perhaps hard to describe the emotions of people with the war ended. Everybody wanted out. Science, in the war years, had been a *fantastic* thing; it had been three fantastic things! It had invented the proximity fuse for anti-aircraft warfare; it had invented radar; and had invented the atomic bomb. At that point the public reaction and the Washington reaction was that scientists were wonderful. Physics, chemistry and mathematics were going to come in for enormous growth. The computer was just being seen over the horizon and, to be very frank, everybody wanted to get back home or to earn a Ph.D. if he did not have one. If he had a job in X university, he wanted to get a better job in Y university. If he had a job in Z university and wanted to be in industry, he wanted to get a job in the best industry at the highest possible pay. The GIs were counting their points as people count beads, for when they had enough points they would be retired from military service.

It was in this atmosphere that I took over. I recall with amazement my own naiveté, for sometime in the Fall of that year I actually organized a recruiting session. Representatives from industry and universities and colleges came around and I arranged for Los Alamos people to talk to them about places to get jobs. Well, maybe it was a good idea, but in retrospect I think I must have been pretty dumb. The situation then was, let's say, chaotic. Almost everyone was looking for a job someplace else. A better job, a good job. The people with sound training wanted more, and quite correctly.

Amdist this exodus, at the end of 1945, we were suddenly faced with an approved demand by the Navy to conduct the first of the effects tests of nuclear weapons in the Bikini Lagoon area. Well, this was a fantastic task. You have no idea of the difficulty. We knew how to make a bomb and fit it in an airplane; that could be done and the bomb could be dropped. But the instrumentation problem was beyond belief; the logistics were fantastic. We'd never done anything like this and I had a group of people, not all of whom felt the greatest loyalty to me or to the project, or to anything else except themselves. Well, we did manage to get the Bikini operation off successfully in the Spring of '46, under a strange set of circumstances: it kept being postponed. Our director would normally have liked this in order to have more time to get things arranged. But it was always postponed at a time when there wasn't *time* to get anything else done. We never got Washington to understand that. But at least Bikini gave us an objective, a challenge, and there's nothing like a task to pull yourself together. So we did.

At the same time I'd like to point out another difficulty that the laboratory faced. Nobody knew what the future was going to be, strange as that may seem. Under whose control was nuclear energy to be? One school of thought, primarily manned by civilians, said that atomic energy was so important, both for potential power uses and even more so for military uses, that it must be under the control of civilians. You can imagine, of course, what the reaction from the military was: atomic energy is so important for military uses that it must *never* be under control of civilians! Until the Atomic Energy Act was passed, the Manhattan District, under General Groves, remained in charge of the operation. I would like to digress here just a moment on the subject of General Groves. He has been almost reviled. He was an extraordinary man with an almost *uncommon* faculty for saying the wrong thing to a scientist. Mostly, I think he was trying to be funny, but he just couldn't pick the right brand of humor. The result was that very few people liked him and many people distrusted him strongly. He was a man without great scientific training; in fact, he was primarily an engineer. However, he listened to the right people. Almost invariably he got a job done. He heard all kinds of advice, but he took the right advice, and he got the task accomplished. He was a remarkable man, not a very loveable man or even a likeable man, but an extraordinary man. He supported me when I desperately needed it, when in 1946 he started building the first permanent housing at Los Alamos.

At any rate, General Groves was working in Washington and I was in charge of Los Alamos. My major problem after Bikini was, bluntly, to pull together a loyal staff, a staff who believed that the laboratory *should* be there and in general agreed what it should do, what kind of laboratory it should be, and what its responsibilities to the country should be. You see, there was one school of thought that said that Los Alamos should be deserted. Put a fence around it, everybody go away, leave it as a monument of man's inhumanity to man. The military, of course, very strongly believed that that was wrong. In their point of view, bigger and better atomic bombs were the obvious order of the day. This debate as to what Los Alamos should do was an extremely important one, and of course I had to be right in the middle of it. If I was going to run a laboratory, I had to have some beliefs of my own. I couldn't ask anybody for direction; I wasn't going to get it! I had to devise a program in which *I* believed and in which I thought I could persuade others to believe.

It was well-known that there were lots of things that had to be done, or certainly ought to be done, in terms of better utilization of materials, better yield-to-weight ratios. There is a basic problem of any device: it has to be delivered; it is no good in your cellar. Its size and its weight interact directly with the delivery vehicle, with the delivery technology. Those things had to be done. Other little bits and pieces, very important bits and pieces that we knew about but simply hadn't had time to explore, were staring at us from behind the scenes, among them the fact that, with the nuclear explosion, one had for the first time the possibility of bringing about the fusion reaction in deuterium: the so-called 'super bomb.' We don't call it that anymore, but hydrogen bomb, fusion bomb, or whatever you wish. The fission bomb opened up a regime of temperature and pressure which were sufficiently in the ballpark, let's say, of those places in the universe where thermonuclear reactions do take place. One could now say, maybe! You see, fission had turned up as a johnny-come-lately. Fusion had been known as a physical phenomenon, let me say, for many years, and people wondered if they could do it on the earth. With a cyclotron, you can bring about fusion reactions. But, no one had seen any way to make this sort of a reaction take place in the bulk because you simply couldn't get the temperature-pressure regimes which were required. But now maybe you could. Maybe. So I felt that these were the sorts of things that had to be done and this country had to do them. I simply did not share and do not share the feeling that if you don't do something nobody else will.

It would have been nice if the Baruch Plan of 1946 had been bought, but it fell like a lead brick. Nobody in the whole world was willing to subscribe to it. Far-seeing, amazing in its general concept, it was twenty-five, or more likely, fifty years ahead of its time. You see, the position that I myself took in that sort of debate was the following. No one makes nuclear weapons or bombs of any sort with any desire to use them. There is no pleasure in using them. In fact, one makes them with a profound desire never to *have* to use them, never to *want* to use them, never to find a *need* to use them. The whole object of the nuclear weapons business has been to put itself out of business. Strange business to be in but it's a fact. And yet, if you asked me in 1945, would we still be making bombs 25 years later, I would have said I don't think so. By that time the major powers would have seen some common sense. But we're still at it. And we have lots of partners in the race. What do you want to give me? Another 25 years, 50 years, 100 years, a thousand years, will we still be making atomic bombs? I simply don't know. I must admit to some personal discouragement; I didn't think it would last this long, but it has. We are beginning to talk to the Russians. Conversations are fragmentary and futile and sometimes silly, but conversations do take place. There is a test ban, of sorts. You don't want to approach that sort of a conference table at the weak side. There's no point in approaching it on the too strong side either, to put it bluntly. Because then the other chap says all right, we aren't going to come to the conference on the weak side either. You have to approach that conference table in a position of relative equality. Then you may be able to find positions where both sides benefit. I simply do not want this country ever to be in a position of coming to a conference table on the weak side.

By September of 1946 a number of things happened. First, there was the Atomic Energy Act of 1946 which remained substantially the same, give or take small variations, for more than 25 years, until the ERDA Act, which abolished it. So at least in 1946 we knew what the Washington organization was going to be. It was a hybrid. The proponents of military control had their say in a Division of Military Application. The proponents of civilian control had a General Advisory Committee and a civilian Commission and civilian managers. So each side got something. Best of all there was a framework. I had a boss, to put it bluntly, a permanent boss, and by the end of '46 the General Advisory Committee and the Commission were known and in January 1947, they made their first official visit to Los Alamos. Something else happened. All through winter, spring, summer, up to September, I was still plagued by the people at Los Alamos not working very hard, maybe not working at all, drawing their pay, because the laboratory policy during the war had been, "We'll pay your way there and we'll pay your way back home, or to wherever your next job may be." As a consequence, some people felt quite free to be rather critical, unhelpful, at the same time drawing their pay. So I finally said, "look, as of September, boys, there'll be no way-home paid anymore." I shook the tree. It had a very desirable effect. I ended up with a residue of around 1400 people who *wanted* to stay at Los Alamos, who wanted to work at Los Alamos, who I suppose in one way shared my own feelings about the role of nuclear weapons and the role of research as a foundation for them. And we started from that point on to build up a laboratory. By 1970, those 1400 had grown to somewhere around 4500.

I had given up any ideas of returning to academic life for very simple personal reasons. You can't persuade anybody to work hard at something if you won't work at it hard yourself, and if I didn't believe in Los Alamos, nobody else would either. And so I had to say, "boys, I'll be director as long as they want me. I will not go back to Stanford or Berkeley," which was a wrench because that's where home was at that time. At any event, there we were. The first thing to do, clearly, was to get at some of the weapons problems that were held over from the war days, ones that we knew about and that could be solved with more work and more effort. General Groves had some very strong ideas about what I'll call the production of nuclear weapons and, of course, Los Alamos was the only place they could be made at that time. People worked very hard keeping up with the schedules. One of the other things that we did was to assert a very strong belief that not all of the laboratory was going to be devoted to making something, or to weapon development. None of us would survive very long if we didn't have a broad foundation, a broad base, of fundamental nuclear research, fundamental chemical, metallurgical, mathematical research, and so on. About half the laboratory's going to do *that* sort of thing, and half the laboratory's going to do *this* sort of thing, and to every extent possible we'll let people trade back and forth so that no man gets all the dirty jobs and no man gets all the nice jobs. Of course, we could not do research across the whole nuclear spectrum; it would be quite impossible with the number of people we had and equally unwise to the taxpayer. So we did our basic, fundamental nuclear research, our materials research, in those areas of chemistry and physics which were peripheral to, or related to, our major objective, which at that time of course had to be weapons. We looked at the properties of uranium, we looked at the properties of plutonium, both chemically and nuclearwise. We didn't forget the thermonuclear reaction, so we made sure we kept our eyes very strongly on all the light elements. Hydrogen, deuterium, lithium, and so on. We confined our research to those areas which were peripheral to our major application efforts.

Well, life went along for three years, to 1948, very pleasantly. That was a time when things had not become regimented or bureaucratized. And I could pick up the phone and call Washington and say, "Hey look, I need some more money." And they would say, "How much?" I would tell them how much I needed and it came! Now, when I say "came" it is to be understood that the laboratory was operated under contract from the AEC by the University of California. One of the things that General Groves did, very wisely in my opinion, was to stay totally clear of civil service. To give you a good reason, civil service bureaucracy is somewhat cumbersome. He believed that the best research is done in the universities. So he asked university contractors to undertake the contractual research responsibilities of these laboratories. And that was indeed the way Los Alamos was run and still is. The AEC has been unique among government operations I might point out. To the best of my knowledge and belief, with minor and now rare exceptions, which ERDA hopefully will continue, the AEC placed its major work with civilian contractors instead of by civil service. At any rate, the University of California was our contractor.

As I said, it was very nice for three years, it was fun, we were making all sorts of progress. The laboratory was growing, exciting things were happening. We had another test in the Pacific in which we made good progress. And then the roof fell in. In 1949 the Russians detonated their first bomb. That was quite a triumph for them. Remember that with all the effort this country mustered on the atomic bomb, it still took us from '42 to '45 or about three years. The greatest technological country in the world, but it took us three years to bring about this task, to make the materials and to make a bomb. It took the Russians not much more. They, surely, had one major start. They knew it *could* be done, which we did *not* know. The great thing that Alamogordo said, and that Hiroshima and Nagasaki said, was that a nuclear explosion was possible. Up to that time one couldn't see any mathematical computational reason why it shouldn't be possible, but it was a regime of temperatures and pressures and behavior that one had never seen in an experimental laboratory before. One couldn't really be sure that the calculations hadn't ignored some impressive fact and that what might have been an explosion would be a poof, or dribble or something. And what Alamogordo showed, of course, was that nuclear explosions can indeed take place, which was a big step forward. The Russians then knew when they started work – and they only started in a major way after the war; they weren't doing much during the war since they had their hands too full in staying alive – that their efforts would be crowned with success if they were just smart. Well, of course, the name of Klaus Fuchs is always bound to occur and how much good Fuchs actually did them. Fuchs was a strange man. I knew him, though not well. A very popular, very reticent bachelor, who was welcome at parties because of his nice manners. He worked very hard; worked very hard for us, for this country. His trouble was that he worked very hard for Russia too. Basically, he hated the Germans bitterly. He had an undying hatred and he simply thought this country was not working hard enough to assist the Russians to defeat the Germans. Well, he was in his own odd way loyal to the United States. He suffered from a double loyalty. But in my own opinion, I doubt if what he got to the Russians through the spy channels was any great help. It may have been some; certainly it wasn't any hindrance. He was a smart man and he doubtless picked up the right things to tell them.

In any event, now we were not alone in the possession of nuclear weapons. Our potential major adversary had them too. Our nice monopoly was gone. We could be as Big Brother-ish as you wish, but we no longer could be Big Brother all by ourselves. And so then the ruckus started. And it was a ruckus that lasted for several years. How do we re-establish our lead over the Russians — until they catch up with us again? The only way anybody could see to do this, aside from what I would call just technological improvements in standard fission weapons, was to find some way of making the fusion reaction work. Now, we hadn't ignored this in spite of some reports to the contrary. But two things had become very clear. Again, some reports to the contrary, what we used to call in a naive way the 'super bomb' probably, probably *would not work*. It was a simple-minded sort of things: we have a fission bomb over here, and some deuterium over here, and you put them side by side, and they go bang. No, you're probably wasting your time on that, although it's almost impossible to prove to the tenth decimal place. And it was becoming pretty clear that you would have to be a lot more sophisticated, very technically smart about fission bombs, before you begin to make a potential fusion bomb. Well, by this time the fat was in the fire. The General Advisory Committee, headed by Robert Oppenheimer, took the point of view, and I would think it was primarily Oppie's point of view, that we had done wrong by the world in developing the fission bomb and that we shouldn't do this again by making the fusion bomb. And if we didn't do it the Russians wouldn't. Well, that was one point of view, a point of view which I disagreed with rather strongly, yet I admit it's a tempting point of view. I simply didn't believe it. On the other hand the military, particularly the Air Force, stimulated by Edward Teller, felt very strongly that if we don't have a superbomb next week the world is going to come crashing down around our ears. This was probably a gross exaggeration too. Well, Edward and I simply disagreed. We disagreed quite definitely on what the best way to go about *trying* to make a fusion weapon would be. What had to be done and how this could be a coherent, logical, reasonable program that didn't fritter away effort in useless and failing experiments. Many, many, *many* calculations had to be done before we would even know how to make a sensible experiment. Ed was much more of the feeling that one should *do* something, do anything, but *do* it, and do it in a hurry. I just couldn't agree. So we somewhat parted company at this point. We're still friendly, but we didn't really agree.

Eventually, in 1951, a bright idea emerged jointly from Stan Ulam, a very well known mathematician, and Edward Teller. It's hard to say about ideas. The day, of course, is long gone in group research when one man sitting in one lab with his own string and sealing wax mades a great invention. The way things go in a laboratory is that somebody sits around the coffee table with somebody else and says, "Hey, I wonder if this would work?" Then some other chap says, "Ah, no, that won't work, but you've given *me* an idea. I wonder if *this* would work." This constant interplay between people, each one coming up with an unworkable idea, let's say, but fertilizing an idea in somebody else's mind is terribly important. Talk goes back and forth and finally somebody says, "Hey, maybe if we try it this way, it might work." And then the way is found. Well, who invented it? No single individual probably invented it, but the patent is at least in the name of Teller and Ulam. There is a patent by the way! For all the good that is! Well, this idea was tested in the Mike shot in 1952. There were some prior tests which were relevant to it and they too were successful. Mike was technically an enormous success. Physically, it was a totally useless device, but it showed very clearly in what direction we had to go to make a weapon, and that was done very rapidly in the following years.

In the laboratory we faced another sort of crucial question. Here we had another jump. We had gone from chemical, pre-World War II explosives to nuclear explosives by 1945, a thousand times more effective. We'd now gone from a few kilotons to megatons, another factor of a thousand. No one saw or sees any way to go another factor of a thousand. One can make these weapons a little better, one can make x megatons or $x + 2$ megatons, but nobody sees how to go from megatons to begatons, high explosive equivalent. It isn't even clear anybody *wants* to. If you get a bomb so big and set it off in your back yard, and that's all you have to do, it takes everybody else with it. So you don't really care if the bomb's much bigger than you have now. You wouldn't know quite what to do with them. Furthermore, it was perfectly clear the Russians were eventually going to get it.

Let me tell you one amusing incident. During the war there was a group of people working at the laboratory known as the British Mission. They were mostly theoreticians, some experimentalists, technical physicists and so on, and special chemists – an extraordinary, brilliant group of people. And they were right in the thick of things. Oppenheimer did not believe in compartmentalizing information, thank Heaven. I copied him, and I didn't believe in it either. So the British knew everything that we knew up to about 1946, when the Mission went home and then the technical contact stopped. Sometime after the Mike shot, contact with the British had been re-established on purely fission things. The British were allies, after all, a small, poor country. And so we had established some degree of technical assistance in this particular field.

The discussions went on partly in England and partly here. It soon became perfectly clear that the British had something else in mind besides the conventional implosion weapon. They hinted around about this and I would have made a small bet that what we were talking about was the same idea that formed the basis of Mike. So I asked if they would be willing to disclose the weight, size and shape of this thing and just what it looked like from outside. They were, and did so – just what you could see. I then hastily asked for a recess because it was perfectly obvious they were describing something that we had gone through great pains in inventing by Teller and Ulam and others. And they'd invented it too. Whether or not there was technical leakage was never learned. In due course the Russians discovered it and the Chinese discovered it. So it's not something that if we hadn't done it, it would never have happened. It would have happened anyway, and I'm glad that we did it first.

The laboratory was at sort of a crossroads. There were many more things to do in nuclear weapons, both fission and fusion, that were of the greatest difficulty. It required the greatest technical competence, the greatest devotion, for a difficult, unpleasant, demanding task. I said it was unpleasant. Nobody *likes* atomic bombs; I hate them. I think you *hate* them; we all hate them. But it has to be done and it had to be done. Much as I look forward some day to an international agreement which banishes the damned things, I didn't want my laboratory to wake up some morning and find it was out of a job. Twenty years or more could be spent refining, modifying, making better, doing all kinds of things with fusion type weapons, making them more effective death carriers. By that time, it was clear things were changing in the technology of weapons delivery. In World War II the B-29 bomber was *the* thing. Incidentally, that's one of the reasons things were sized as they were. But the missile had begun to appear, or would shortly appear after Sputnik, and the missile was going to replace the airplane in many ways as the prime delivery system. And the problem was to coordinate the warhead with the delivery system. A whole new set of technical problems appeared: 'clean' bombs; as you know there was a lot of talk about that. Clean bombs, bombs with specialized effects, bombs with specialized shapes to adapt themselves to the peculiar shapes that intercontinental ballistic missiles like to have, things of that sort. But those things could easily be done, and you'd have the same sort of technological refinement and sophistication that followed 1945, and up to that time was still going on.

Well, here you have a new problem. You have a very difficult task, to be sure, but if somebody thinks his job is soon over and the place is going to be shut down when atomic bombs are through, he'll think to go out and get a good job *now* and not wait 'till he gets his pink slip. So it became very important for *me*, just because of the importance of the weapon program to start devoting a considerable fraction of the laboratory's efforts to other applications of nuclear energy, with the special competence the laboratory had developed in weapon work and in other research which we thought was relevant. And that is indeed what we did. The weapon work, instead of being a major fraction of the laboratory effort, was gradually reduced through the latter part of the '60's to perhaps 25 to 30% of the laboratory effort. Another 25–30% was fundamental research, and other 25 or 30% was devoted to other applications of nuclear energy, particularly to peaceful uses. After all, as we know to our cost today, if you can invent better power sources, you have as good a weapon *vis a vis* the rest of the world as you had with an atomic bomb, and perhaps better.

And we did many other things: nuclear medicine, controlled thermonuclear reactions, and so on. These were done rather deliberately and with the object of making sure that the laboratory was a permanent, viable institution, viable for the tasks of today, and viable ten or twenty years from now with the tasks of tomorrow. We started a reactor program. The laboratory had always had a little reactor work. Up until about a year ago it had the oldest operating reactor in the world – an old water boiler. That thing ran for twenty-five years or more and did lots of good research. We made the first reactor that utilized enriched material. We made the first plutonium reactor, called Clementine, and it was the first reactor that operated upon a fast neutron spectrum. I only pause to note that today it would take you ten billion dollars and fifty thousand volumes of environmental reports and nobody'd let you do it anyway, but we just *did* it. And it worked beautifully; we finally shut it down when it became clear that some of the components were a little tired. Then we invented another nice type of reactor, which unfortunately didn't sell then; it may never sell, but it was absolutely self-contained. It would go in a hole in the ground about a meter in diameter. This reactor developed steam and turned itself off or on as the power demand required. It could be used for places like radar posts in the Arctic, at that time; now, possibly for remote drilling stations there. Well, this was a *power* source. It was a beautiful thing, done by one of our eminent reactor people, Percival King.

Elsewhere we developed our expertise in plutonium. Plutonium is a major component of nuclear weapons, fission or fusion, and that's why you have places like Hanford. It was clear to anybody who was looking at the nuclear power business that you should find some way to burn uranium 238, since uranium 235 was only 0.7 of one percent in the ore you mine. Thus, the dent you'll make upon the power needs of the world is over a hundred times too small. So you should find some way to burn 238, and indeed you know how to do it. It's called a breeder and it's been called a breeder for many years; but unfortunately we don't have one yet. Now, breeders are tough to make; there's no doubt about that. In a breeder reactor you convert some of the uranium 238 to plutonium. Then you burn that plutonium. One can burn plutonium directly in the reactor, which in this case is a breeder, or can separate it from the material and then make a plutonium reactor. That was our first reactor, a type of plutonium burning reactor. But it was no power reactor; it was a research reactor. A power reactor is a very different breed of cat. So we decided to find some way to make power-producing plutonium reactor elements. That was a tough one. Unfortunately, another set of problems hit us about that time and the work eventually got closed down. I'm sure that the people that worked on it are long retired. I wish they had succeeded. We were on the trail of using liquid plutonium as a fuel and while that certainly has its own set of problems, it nevertheless is the way I bet we end up going. So Los Alamos went into this sort of reactor work rather heavily.

The same thing is true for controlled thermonuclear reactions. CTR work starts off on the premise that if you can make a fusion bomb, and you can make a fission bomb, and you can make a fission reactor, why can't you make a fusion reactor? What's so hard about it? Well, the problem is still unsolved. After twenty-five years, in spite of the most extraordinary efforts by the Russians, British, French, everybody else, and particularly by the United States, we still have something like a factor of a hundred between the energy you put into a fusion reactor experiment and what you get out. That of course gets you nowhere fast. I wish I could say that you could see signs of this number improving greatly across the world. Machines get bigger and bigger, but real progress seems very slow. I hope Nature isn't kidding us here; I don't suppose she really is. The energy is there, but it certainly is very difficult to extract and it's going to be very expensive to extract.

I wish people would stop talking about energy in this sense. There's as much energy in the world as there ever was; you can't destroy it. What you do, of course, is to make it less and less available. And when you burn up that valuable gasoline in your car, all you do is convert it to heat which warms up the atmosphere a little bit, which isn't the slightest use to anyone. But the energy is still there; you haven't lost it. It's just that we are using up the *available* sources and having to go to less and less available sources, and those poorer available sources become expensive. Plants cost money to build and money costs interest and you, the consumer, pay for it. The interest charges can kill you. The fuel may be infinitely available for a controlled fusion reactor, cheap, but if the plant costs you ten billion dollars, you're going to hurt in those interest charges! I wish I could be confident that this would get solved; I used to be more confident than I am now. But twenty-five years have gone by now with the smartest people in the business worrying about this and they didn't find the answer. Still, it may be found by some bright young graduate student tomorrow. In the science business it's no use to say nobody will ever discover it. Fusion and fission were discovered in the thirties so don't ever say of anything, "it can't happen." The right idea may not have appeared, but old ideas are being resurrected, even now. It amuses me because this example was discussed about 1960. One of the ways of getting fusion power is to take a nice clean nuclear bomb, set it off in a great big underground tank, fill a tank nearly full of water pipes and so on, run cool water in here and steam comes out there. That may be one of the practical ways of getting fusion power. I say it not entirely seriously, but we surely could do it. I'd worry about what the cost might be, but it could be done.

Another thing that happened in those days was that the Russians surprised us with Sputnik. And that was one of the biggest surprises that this country ever had. A blow to our confidence if you wish. The Russians suddenly made a satellite. Furthermore, if you can do that you can make a missile which goes from here to there, and that is an obvious threat. If you can get something to go around the earth with that sort of precision, you can always get something which goes accurately to a place *on* earth. And that gives you the creeping shivers. So the United States, in one of its most cataclysmic scientific changes, encouraged school children to learn physics and chemistry and binary systems of numbers and everything else. We produced many Ph.D.s, many M.A.s, and many graduate students. Of course now we don't feel quite the same way and sometimes they walk the streets, jobless.

But that again is a problem of our country and I will illustrate it with my final point. Los Alamos being Los Alamos, we said, "Well, how can we help in this?" The place where we thought we might help first of all was to devise a nuclear-powered rocket motor for an intercontinental ballistic missile. Believe it or not, in those days there wasn't such a thing. There wasn't any intercontinental ballistic missile; the chemists hadn't even bothered, the Air Force hadn't bothered; nobody had bothered. The chemists were scratching their head a bit wondering if they could. Now, however, the Air Force was jumping up and down for obvious reasons. So we said, "Look, we think the chemists can do it. But in case they can't, we'll see if we can produce a nuclear power plant for an intercontinental ballistic missile." So we started bravely out along that line and by the time we had gotten a year or so along it the chemists were clearly successful. They were going to get it done ahead of anybody and at one-tenth of the cost. So we backed off that particular program. But then, of course, we had to keep up with the Russians in space and we started to look at the rocket equations. There is one thing that nuclear power can do for rocket propulsion, space rocket propulsion, that you can't do with ordinary chemical combustion. You can choose the molecular weight of the gas which you eject through the nozzle. The chemical motors, of course *have* to use what they're burning, what they're combusting; they're stuck with it. Those molecular weights are generally fairly high. Not too high, but unpleasantly so. And the '*goodness*' of a rocket motor, which is called the specific impulse, depends upon the reciprocal of the square root of the molecular weight of the propellant gas. It's like miles per gallon. (It isn't, but it's something like that.) In a nuclear-powered rocket you may choose your rocket fuel, and since you try for the smallest molecular weight you can, you pick hydrogen. That sounds easy; I wish it were. Now we worked very hard for some years, with many of the taxpayer's dollars, developing a graphite-based, hydrogen-propelled nuclear type of rocket. It was the Rover program. And then again a small frailty in the way this country runs research appeared. What we started to do was to demonstrate the feasibility of a nuclear rocket engine, not its use. But this was picked up by Washington and we were told to get these things into space no matter what. Now all kinds of things happened. At one point we were spending 70 to 80 million dollars a year on this program. Today that program is as dead as a doornail. Why? Well, space is not dead, but space isn't very exciting anymore. We spent too much too soon, and instead of having a good long-range program we went in like a firecracker and came out too abruptly. At one point we had to fire quite a number of people just because of the cancellation of this program.

This country does not always know how to run its long-range programs. The basic problem is this: major programs today, the nuclear reactor, breeder reactors, controlled thermonuclear fusion programs, and the like, take years and years and years. I'm speaking of decades. But the professional lifetime of some manager in Washington, if he's lucky, is possibly five years. And so what turns out to be one man's meat may be another man's poison in some types of programs. And no man is ever held to account for his errors. When mistakes are made and discovered in the reactor business, the chances are good that the individual who made them is long gone. What is one going to do about it? Programs last so long, by nature, that the man who starts the reactor research sometimes doesn't live to finish it. It used to be a sort of standing joke that in our nuclear rocket work we felt similar to the people who built the cathedrals in Europe: they were started by the grandparents and finished by the grandchildren. The last thing that I managed to accomplish before I retired was to get Washington's approval to build a very large, half-mile long accelerator for the production of some nuclear particles, pions, and a so-called meson factory, which is now running and doing useful research. And you say, what's that for? It's not for bombs, it's not for energy, it's just plain good physics, and the argument for doing plain, good nuclear physics has to be what it always was. You've got to look under every stone and see what might be there. If you hadn't looked under certain stones about neutrons versus uranium in 1938–39, you'd never have found fission. I don't think that this accelerator is very likely to do more than produce good physics, good understanding of sub-nuclear physics, sub-nuclear particles, medical-use discoveries to deal with malignancies because of certain characteristic ways mesons react with tissue. You simply cannot let the country leave stones unturned. There may not be anything there, but suppose there is. You'd better find it.

BIOGRAPHICAL NOTES

NORRIS BRADBURY received his doctorate in physics in 1932 from the University of California at Berkeley. With a National Research Council Fellowship he then spent two years doing research at the Massachusetts Institute of Technology, after which he joined the faculty at Stanford University, becoming a full professor in 1943. In 1941 he received a commission in the Naval Reserve and served at the Naval Proving Ground, Dahlgren, Virginia, until the summer of 1944. During the final year of the war he was in charge of the impolsion field-test program at Los Alamos and responsible for the assembly of all non-nuclear components of the plutonium bomb.

On Oppenheimer's recommendation, Dr. Bradbury was released from active duty in October 1945 and became his successor as director of the Los Alamos Laboratory. He was, Bradbury believed, just an interim director of a laboratory that might well soon close. However, by his retirement twenty-five years later, the wartime fenced-in mesa had changed to a thriving open city, with a laboratory impressive for its size, range of interests, and significant accomplishments.

Dr. Bradbury is a member of the National Academy of Sciences, has served on the Air Force's Scientific Advisory Board and the Scientific Advisory Committee in the Office of Defense Mobilization, and has received the Department of Defense's Distinguished Public Service Medal.

BERNICE BRODE attended Occidental College, the University of California's Southern Branch (now UCLA), and received both her bachelor's and master's degrees from the University of California at Berkeley. As the wife of Berkeley physics professor Robert Brode, she shared his fellowships and visiting positions at Princeton, London, Cambridge, Manchester, MIT, and in Washington, D.C. In 1942–1943, Mrs. Brode taught English at the Russian Military Mission in Washington and, as she describes in her lecture, in 1943–1945 worked as a computer at Los Alamos. A frequent traveler abroad, she served as assistant to the director of the University of California's Education Abroad Program during an extended residence in London. Both these travels and her association with the scientific community have served as the subjects of a number of published articles.

177

L. Badash, J. O. Hirschfelder and H. P. Broida (eds.), Reminiscences of Los Alamos
1943–1945, 177–180.
Copyright © 1980 by D. Reidel Publishing Company.

JOHN H. DUDLEY, a graduate of West Point, Massachusetts Institute of Technology, and the Army War College, served in the Engineer Corps. In 1942–1943 he played a key role in selecting a site for the secret weapons laboratory and was responsible for initial construction there.

Before his retirement in 1960 as a Brigadier General, he was academic head of the U.S. Army Engineer School. He is now professor of civil engineering at California State University, Long Beach.

LAURA FERMI became famous in her own right with the publication of *Atoms in the Family* in 1954, the same year in which her husband died. They were married in 1928, when Enrico Fermi was a professor in Rome. They fled Fascist Italy in 1938, when the award of the Nobel Prize for Physics to Enrico offered an opportunity to go abroad. His work subsequently took them to Columbia University, the University of Chicago, Los Alamos, and back to Chicago after the war.

Mrs. Fermi's warm and interesting book about the experiences of her family was followed by *Atoms for the World* (1957), written while a consultant to the Atomic Energy Commission; *Mussolini* (1961), a biography whose research was supported by a Guggenheim Fellowship; and *Illustrious Immigrants* (1968), the story of those who left Europe at the same time as the Fermis.

RICHARD P. FEYNMAN, Nobel laureate for his work in quantum electrodynamics, is well known also for his skill in explaining science to students and the layman, and the sense of humor he brings to this role. As his lecture indicates, this latter characteristic was well developed by the time he received his Ph.D. in theoretical physics from Princeton University in 1942, and was given scope for further refinement at Los Alamos between 1942 and 1945. From 1945 to 1951 he was a faculty member at Cornell University, and in 1951 became professor of physics at the California Institute of Technology. He is a foreign member of The Royal Society (London).

JOSEPH O. HIRSCHFELDER received his Ph.D. degree from Princeton in 1936 and subsequently held several academic positions at the University of Wisconsin. For the first part of the war he was group leader in the National Defense Research Committee, where he developed the systems for predicting the interior ballistics for both guns and rockets which were adopted by the armed services. At Los Alamos he was a group leader in both the Ordnance and the Theoretical Divisions, with his most notable accomplishment being the prediction of the effects of the atomic bomb. He was chief phenomenologist for the 1946 Bikini bomb tests, and chairman of the board of editors of *The Effects of Atomic Weapons*, which was published in 1949. Since 1946 he has been professor of chemistry at the University of Wisconsin, and for the past fifteen years director of its Theoretical Chemistry Institute. He is a member of the Norwegian Royal Society and the United States National Academy of Sciences, and has received the National Medal of Science in 1976 from President Ford, the American Chemical Society's Debye Award, the Combustion Institute's Egerton Gold Medal, and the honorary D.Sc. degree from Marquette University.

GEORGE B. KISTIAKOWSKY was born in Ukraine, Russia in 1900, educated there and in Germany, receiving his doctorate in chemistry from the University of Berlin in 1925. During the next five years he was at Princeton University, first on a fellowship and then as a member of the staff. In 1930 he moved to Harvard University, where he became professor of chemistry in 1938 and emeritus professor in 1971. At Los Alamos he was chief of the Explosives Division. After the war he was a member of the President's Science Advisory Committee from 1957 to 1964, served as Special Assistant to the President for Science and Technology during the last year and a half of Eisenhower's administration, and was a member of the General Advisory Committee to the U.S. Arms Control and Disarmament Agency, 1962–1969.

Professor Kistiakowsky was vice president of the National Academy of Sciences from 1965 to 1973, and first chairman of the NAS Committee on Science and Public Policy, 1962–1965. He is a foreign member of The Royal Society (London), and the recipient of many awards, including the Medal for Merit, Medal of Freedom, and National Medal of Science from three Presidents of the U.S.A., as well as the Willard Gibbs Medal, Peter Debye Award, Theodore William Richards Medal, and the Priestley Medal.

JOHN H. MANLEY received his doctorate in physics from the University of Michigan in 1934, was a lecturer at Columbia University from 1934 to 1937, and then an associate at the University of Illinois to 1941. In 1942 and 1943 he was attached to the Manhattan Project's Metallurgical Laboratory in Chicago; from 1943 he was at Los Alamos, principally as Oppenheimer's aide in creating the new laboratory. After World War II, Dr. Manley sandwiched the associate directorship of Los Alamos (1947–1951) between professorships at Washington University (St. Louis) and the University of Washington (Seattle). He served also as secretary of the Atomic Energy Commission's General

Advisory Committee, and briefly as deputy director of the AEC's Division of Research. In 1957 he returned to Los Alamos as a research advisor.

EDWIN M. McMILLAN was awarded his doctorate in physics from Princeton University in 1932, whereup he began his lifelong academic career at the University of California at Berkeley. There he was associated with the physics department and the Radiation Laboratory, becoming the latter's director in 1958 upon the death of Ernest Lawrence. Between 1940 and 1945 he was on leave of absence, working successively on radar at MIT, sonar at the Navy laboratory in San Diego, and nuclear weapons at Los Alamos.

In 1951, Professor McMillan shared the Nobel Prize in Chemistry with Glenn Seaborg for their work on the transuranium elements, McMillan having discovered (with Phillip Abelson) the element neptunium. In 1963, he shared the Atoms for Peace Award with V. I. Veksler for their independent conceptions of the principle of phase stability, which is essential to the design of high energy particle accelerators. He is a member of the National Academy of Sciences, served on the General Advisory Committee of the AEC (1954–1958), and is now an emeritus professor in the University of California.

ELSIE McMILLAN was born in New Haven, Connecticut, where her father was dean of the Yale Medical School. Before her marriage to Edwin McMillan in 1941, she worked as a kindergarten teacher, radio performer, and occupational therapist. Since their residence in Los Alamos they have lived in the Berkeley area, where Mrs. McMillan has been active in writing and lecturing, as well as in her role as a wife and mother.

INDEX

STUDIES IN THE HISTORY
OF MODERN SCIENCE

Editors:

ROBERT S. COHEN (Boston University)
ERWIN N. HIEBERT (Harvard University)
EVERETT I. MENDELSOHN (Harvard University)